WATER SUPPLY AND DEMAND MANAGEMENT IN THE GALÁPAGOS: A CASE STUDY OF SANTA CRUZ ISLAND

DISSERTATION

Submitted in fulfilment of the requirements of
the Board for Doctorates of Delft University of Technology
and
of the Academic Board of the UNESCO-IHE
Institute for Water Education
for
the Degree of DOCTOR
to be defended in public on
Thursday September 28, 2017, at 15:00 hours
In Delft, the Netherlands

by

Maria Fernanda REYES PEREZ

Master of Science in Environmental and Energy Management,
Twente University, Enschede- The Netherlands

born in Quito, Ecuador

CRC Press
Taylor & Francis Group
Boca Raton London New York

CRC Press is an imprint of the
Taylor & Francis Group, an **informa** business
A BALKEMA BOOK

This dissertation has been approved by the
promotor: Prof. Dr. M. D. Kennedy
copromotor: Dr. N. Trifunovic

Composition of the Doctoral Committee:

Chairman	Rector Magnificus, Delft University of Technology
Vice-Chairman	Rector UNESCO-IHE
Prof. Dr. M. Kennedy	UNESCO-IHE / Delft University of Technology, promoter
Dr. N. Trifunovic	UNESCO-IHE / Supervisor

Independent members:
Prof. Dr. C. Mena	Universidad San Francisco de Quito/Galápagos Science Center
Dr. P. Gikas	Technical University of Crete
Prof. Dr. D. Butler	University of Exeter
Prof. Dr. L. Rietveld	Delft University of Technology
Prof. Dr. M. McClain	Delft University of Technology / UNESCO-IHE, reserve member

This research was conducted under the auspices of the Graduate School for Socio-Economic and Natural Sciences of the Environment (SENSE)

First issued in hardback 2018

CRC Press/Balkema is an imprint of the Taylor & Francis Group, an informal business
© 2017, M. F. Reyes Perez

Published by:
CRC Press/Balkema
Schipholweg 107C, 2316 XC Leiden, The Netherlands
e-mail: Pub.NL@taylorandfrancis.com
www.crcpress.com – www.taylorandfrancis.com
ISBN 13: 978-1-138-37319-8 (hbk)
ISBN 13: 978-0-8153-7247-9 (pbk)

"Emancipate yourselves from mental slavery, none but ourselves can free our minds."

- Bob Marley

A mi familia con todo mi amor,

Voor mijn Klein met lief.

Acknowledgements

This thesis is the result of five years of hard work in Santa Cruz Island, located in the wonderful Archipelago of Galápagos, and Delft. Every single person I encountered along this journey, has somehow contributed to this piece of work. Therefore, I would like to name each one of them and thank them (I hope not to forget anyone).

First of all, I would like to start by thanking my sponsor SENESCYT (Secretaría de Educación, Ciencia Tecnología e Innovación), who contributed financially so I could carry out my PhD. Then, many special thanks to my promoter and Professor Maria Kennedy, who has always been very kind, supportive and understanding, who always helped me to improve every chapter in every possible way. I thank you very much for your criticism because it made me see things that, after some time, I could not recognize anymore. I thank you deeply for all the feedback and ideas, as well as for the financial support towards the end of the study. Thank you Maria!

My sincere gratitude to my mentor and friend, Nemanja Trifunovic, who is not only an expert in the field of water supply and distribution, but is also a very wise, patient and enthusiastic human being, always carrying a smile on his face and having positive comments about any circumstance (really admirable!). Thanks Nemanja for your help, your words of support, your understanding in the moments of crisis and above all for your patience and contribution to this research. I hope we can enjoy in the future a fresh grilled tuna on the beautiful island of Santa Cruz. Also, thanks for your help on my writing, I know it was very hectic all the repetitions, as well as the elaborated and run-on sentences☺.

Also, to Saroj Sharma, my second mentor who had always a positive comment and word of motivation in the joint meetings. Thanks also for all the positive comments and fast revision of manuscripts, they really helped me to organize my ideas. Also, to Chantal and Bianca, who were always so efficient in any administration matter I needed. To Jolanda Boots, the best PhD admission officer there can be. To my MSc. student Diah Prameshwari for the contribution on Chapter 6 and many special thanks to my other MSc. student Aleksandar Petricic, for all your contribution on Chapter 7, your hard work was admirable and helped me a lot, especially on the second or third analysis! Thanks a lot!

This study would not have been possible without the contribution of all the people in Santa Cruz Island. To Delio Sarango, the chief of the Department of Potable Water of Santa Cruz, so many thanks for all the information provided, for giving so much of your time to this research, for always being reachable through whatsapp and answering questions along five years of work, as well as lending me the 18 water meters, equipment, transportation and so many other things. Also, Delia and Sandra, who were always willing to contribute with data and helped me find the families to install the water meters. Thanks to the personnel who helped on the installation and un-installation of the meters: Don Efren Valle, Vinicio Gaona, Jose María Masaquiza , Don Juan Pacheco and Lorena Intriago. Also, thanks to Daniel Proano for being part of the project and providing ideas for the interviews. To Eva Anagono, Lorena, Joanna and Mariano for helping me carry out the surveys, which was a lot of hard work. In the Dirección del Parque Nacional Galápagos, thanks to Wacho Tapia with all the administration process for my research stay, as well as all the advices. Thanks also to Galo Quezada, who also facilitated my stay in the islands. Thanks to Noemi d'Ozouville for the contacts provided. Also many thanks to Carlos Mena, Steve Walsh and all the staff of the Galapagos Science Center, which were always willing to help and contribute. To Tito Guerra from SENAGUA for the data provided. To the USFQ, my local university and the people in the Environmental Engineering Department: Rene Parra, Rodny Penafiel and Valeria Ochoa. Lastly in Galapagos, thanks to the friends there who helped me as well: Mayrita (la major masajista del mundo), Kari, Osamu, Sebas, Vale... you made me have a wonderful time there!

To all my friends in Ecuador, who were always there to support me and encourage me to continue and do not give up. Goshguis Vivi y Benji, my best friends since elementary school, who are always there, no matter how long it is that we don't speak, it feels like not a day has passed by. Many special thanks to Vivi Lopez for your creation for the cover of this book and all the time spent! After so many discussions, we finally got a beautiful version. To Dani Barriga (polla), Cris Viteri (alias Fago), Andres Becdach (Giito), Pablo Morales (Peladito) y Esteban Ruiz (Ivy), my closest friends in life, who I miss every day and miss our university "crazy" years. Also, llamita Caro Cordovez and Cris Gomezjurado, even though we met few years ago, I know that I can always count on you girls.

To the gang in Delft: Vero, Pato, Jessi, Luisita, Erika, Alida, Juliette, Mauri, Yared, Aki, Mark, Pablo, Neiler and Miguel (veci). So many memories with all of you, who made it this journey, somehow easier. Also to the amigosss/running/partying group, my family here: Mohaned (Mohi): thank you for your patience with my Dutch grammar☺. I will never forget the 'onafhankelijkheid' word, which I had to repeat close to 50 times so could pronounce it properly! Berend, so many thanks for the talks, words of comfort, and all the fun. I love your dance steps and I hope you can replicate them when we travel all together to Ecuador and Colombia. Also to Can, who became member of the group later, but still very special to all of us. Thank you for always being available for some shoes advice (haha). Thanks amigos for all the parties, 'despacito' dancing, all our dinners and tequilas, we always had such a great time!! I am really going to miss you all! Very special thanks to Juanca and Motassem, who helped me with hydraulics, calculations and the EPANET model. You both were my little angels who helped in my moments of despair (hahaha). To my Ecuadorian friends Gaby and Saira, who always made me feel at home!

To my girls, Angie and Nata (my paranynmph): without you girls, my life in Delft would have been boring, but you really spiced it up. All the parties, the dinners and your friendship was simply something that made me want to stay. I will never forget our trip to Terschelling and all the anecdotes! I will miss you girls so much, all our talks, coffees, carrot cakes, dinners, trips, kapsalons and all the after-parties at Kobis Grill, please don't forget our anthem: Mi No Lob. You girls are simply the best and I can't thank you enough for your friendship and for becoming my sisters here. Thanks also for all the personal training and routines at the gym, every day was fun with bodypump, spinning, Angie's weight programs, running and 10 K's competitions.

To the love of my life, Peter, who is the motor of my life: without you I don't think I would have been able to finish. Thanks for all your unconditional support, for all your love, for all your care and your HELP. Thanks for always listening to me and always finding the right words to comfort me. Thanks for the Dutch lessons and correcting my schrijfopdrachten and many special thanks for the translation of the summaries, which, I know were very hectic!! You are the most amazing man I have ever met and I am so happy you have become part of my life; you are my TWO. I love you with all my heart! This is just the beginning of an amazing trip together. Ik hou van jou mijn kleintje!

Also thanks to my Dutch family for their support and care: Tineke, Ap, Marjan, Daan, Isabelle and Jasmijn. Thank you for always making me feel I was at home, and being so caring!

Last, but not least, to all my family. To my mother Charo, who is the most remarkable woman on the planet, smart and beautiful, you have always been my role model. To my brother, Goyo, who always gave me free therapy sessions in my moments of despair and my other paranymph. Thank you for checking my thesis ☺. To my father, who, despite his personal journey, he made me feel he was always proud of me. To Angie, Juani, Samirula, Mami Ro and all the aunts, uncles and cousins who welcomed me every time I went to Ecuador with a party, a cafecito, or a delicious cangrejada. Also, thanks to God who gave me this opportunity and put the right people at the right time. I can't be more thankful for this amazing 5 year journey!!

Gracias a Dios y a la vida por todo lo que me han brindado, solo puedo decir que me siento bendecida!

List of Figures

List of Tables

Table of Contents

Summary

The Galápagos Islands, a province of the Republic of Ecuador, is a volcanic archipelago of significant ecological importance. For centuries, water resources on the islands have been perceived as scarce and consequently, water issues have been recognized as an important and urgent matter. As in many other tourist islands, water resources have been severely threatened by the expansion of the tourist industry. In addition, data regarding water supply and demand are scarce since the vast research carried out on these islands focuses mainly to the conservation of endemic species, putting aside the impact of human activities on water resources.

Santa Cruz Island, the hub of tourism of the Galápagos Archipelago, is experiencing significant challenges regarding water quantity and quality. There are no permanent freshwater resources in this island. The municipal system supplies only (untreated) brackish ground water (800-1200 mg/l of Chloride) intermittently. Water scarcity is felt mostly in the town of Puerto Ayora, which is the centre of tourism, where the average supply is three hours per day. In Puerto Ayora, there are no water meters and the water tariff structure is fixed. The municipal supply system has not been able to cope with the current growth in tourism and local population (7% annual growth of tourism and 3.3% annual growth of local population). Non-revenue water ranging from 35% (this thesis) to 70% (previous studies) on the island, poor maintenance and old piping systems adds also to the provision of an erratic service within this island.

The research presented in this thesis focuses on water supply and water demand management on Santa Cruz Island. Firstly, water supply was analysed in order to identify the issues contributing to the current situation regarding the intermittency of water supply. Results showed three sources of supply: (1) brackish ground/crevice water distributed by the municipal supply, (2) bottled-desalinated groundwater produced by small private companies and (3) brackish ground/crevice water privately extracted and sold by water trucks. From this study, the estimated quantity supplied per capita (including all three sources) is ±370 lpcpd, which is high compared to the capital city Quito (210 lpcpd).

Thereafter, water demand was estimated considering the different demand categories such as domestic households, hotels, restaurants and laundries. Water demand was assessed through approximately 400 surveys (including domestic households, hotels, restaurants and laundries), which were distributed all around the town of Puerto Ayora. The quantification of water demand was performed for each of the three sources previously identified. Results showed that the average per capita demand from the municipal supply was estimated as ±163 lpcpd, which can be considered high for an island with no freshwater resources and intermittent supply for approximately 3 hours per day. Later, in order to verify the specific domestic demand, 18 water meters were installed around the town of Puerto Ayora. In addition, the readings from approximately 300 water meters previously installed on three pilot zones established by the municipality were analysed and compared with the results of the 18 water meters add as part of this study. Both showed very high domestic water consumption, the average obtained from the 18 water meters showed an average demand of 164 lpcpd ± 94 lpcpd of standard deviation. The average domestic demand for the pilot zones were estimated at 182 lpcpd ±31, 195 lpcpd ±80 and 428 lpcpd ±70, for pilot zone 1, pilot zone 2 and pilot zone 3, respectively. In many of the pilot zones outliers were identified with average consumption as high as 4,500 lpcpd, suggesting excessive wastage of water within households and/or informal tourist accommodations, or both.

A prognosis of urban water supply and demand was carried out for the next 30 years, for four different annual growth scenarios. These scenarios were: (1) slow growth- 1% for the local population and tourism, (2) moderate growth- 3% for local population and 4% for tourism, (3) fast growth- 5% for local population and 7% for tourism (current situation) and (4) very fast growth- 7% for local population and 9% for tourism. Results showed that without any intervention (business as usual) water demand coverage will barely reach 50% for the slow growth scenario and 10% for the very fast growth scenario. Consequently, five intervention strategies were developed and evaluated as options to solve current and future water scarcity. The strategies included sustainable options such as reducing per capita water demand, installing water meters, reducing leakage, rainwater harvesting and grey water recycling as well as the installation of a seawater desalination plant on the island. These strategies were assessed using several Key Performance Indicators (KPI's), in terms of water demand coverage with supply, costs and energy use over the next 30 years. Results showed that the intervention strategy involving the installation of a seawater desalination plant is the only strategy that can completely meet the demand in a future 'rapid population growth' scenario, while simultaneously improving water quality (reducing salinity at the tap). However, this is the most expensive and energy intensive solution for the island. The intervention strategy that includes all options, except desalination, will suffice, but only for the 'slow population growth' scenario, which is very unlikely to happen.

Afterwards, a Multi-Criteria Decision Analysis was performed, with the aim of evaluating the five intervention strategies assuming a 'moderate population growth' scenario. The strategies were categorized and evaluated under four criteria: environmental, technical, economic and social, and considering four groups of stakeholders: local decision-makers (governmental authorities), local experts (researchers, academia and environmentalists), domestic end-users and hotels. The proposed strategies were finally ranked based on the different stakeholder's perspectives, providing the preferred strategy considering the selected criteria. Results differed for each stakeholder group: for local decision makers the intervention strategy including the installation of a desalination plant ranked first, while for local experts, domestic end-users and hotels, that strategy was ranked last and they preferred more environmentally friendly options. In addition, a sensitivity analysis was carried out, which showed that the most sensitive criteria are the environmental, technical and social ones, and small changes in the values of their weights may significantly change the ranking of the intervention strategies.

In addition, the water supply network of Santa Cruz was evaluated with EPANET software, aiming to assess the current and future performance under the different growth scenarios, using Demand Driven Analysis (DDA) and Pressure-Driven Analysis (PDA) approaches. Each approach suggested that the current network suffices for 24 hour supply, in terms of quantity of water and pressure available within the network. Likewise, the household storage facilities were evaluated, developing a methodology (using the Emitter Coefficient feature of the software) with the aim of estimating water loss (overflow) from roof tanks for several scenarios of water consumption, leakage and storage tank capacity. Results showed that water losses from roof tanks varied from 5 to 32% of the total water supplied in the town of Puerto Ayora.

The water demand estimated, as well as water losses within households is surprisingly high for an island where there are no permanent freshwater resources and intermittent supply for three hours per day. Most likely, this is a direct consequence of a fixed water tariff structure, which does not provide any incentive to people to save water. Therefore, any future growth in tourism should be limited. If the very fast tourist growth scenario is defined as a governmental target (9% annually), inevitably, a seawater desalination plant will need to be installed. This will solve water quantity and quality issues, and is independent of the amount of rainfall on the island, but many negative environmental impacts may be generated, especially regarding brine disposal, chemical discharge, energy consumption and fuel importation. For the fast and moderate growth scenarios, the proposed sustainable strategies (including leakage reduction, water meter installation, per capita demand reduction, rainwater harvesting and grey water recycling) would be partially sufficient, suggesting that at least for some activities a smaller seawater desalination plant may need to be installed as well. For the slow growth scenario, the proposed intervention strategy comprising sustainable strategies would be sufficient, suggesting that governmental tourism targets should be re-considered. If current growth trends continue, the overexploitation of brackish water from the basal aquifer may increase the salinity of the supplied water, driving the need for a desalination plant. Therefore, the amount of water that is currently consumed on the island needs to be addressed, fixed water tariffs should be abolished and the governmental targets for tourists visiting the island should be re-considered in order to preserve this fragile and unique ecosystem.

Samenvatting

De Galápagos eilanden, een provincie van Equador, is een vulkanische archipel met een wezenlijk ecologisch belang. Dat maakt deze eilandengroep tot een unieke plaats in de wereld. Eeuwenlang heeft men waterschaarste op de eilanden ondervonden en als gevolg daarvan wordt water problematiek als belangrijk en urgent beschouwd. Zoals op vele andere toeristische eilanden zijn waterbronnen ernstig bedreigd door de significante groei van de toeristen industrie.

Bovendien is informatie betreffende water levering en behoefte schaars aangezien uitgebreid onderzoek op deze eilanden zich voornamelijk richt op het behoud van endemische soorten, waardoor het effect van menselijke activiteit op water bronnen naar de achtergrond wordt verplaatst.

Santa Cruz eiland, de toeristische 'hub' van de Galápagos archipel ervaart grote uitdagingen voor wat betreft waterkwaliteit en -kwantiteit. Er zijn geen permanente zoet water bronnen op dit eiland met uitzondering van regenwater. Daarom levert het gemeentelijk systeem, met tussenpozen, alleen onbehandeld brak water.(800-1200 mg/l Chloride)

De grootste uitdagingen zijn voelbaar in Puerto Ayora, het toeristische centrum, waar de gemiddelde levering 3 uur per dag bedraagt. In Puerto Ayora zijn er geen watermeters en is er een vaste tarief structuur.

Het gemeentelijk systeem is niet in staat gebleken om te kunnen gaan met de huidige groei van toerisme en lokale bevolking (7% jaarlijkse groei van toerisme en 3,3% jaarlijkse groei van de lokale bevolking).

'Non-revenue' water variërend van 35% (in deze thesis) tot 70% (eerdere studies), slecht onderhoud en een verouderd leidingsysteem dragen bij aan een instabiele waterservice op het eiland.

Het onderzoek gepresenteerd in deze thesis concentreert zich op water levering en water behoefte management. Als eerste is de water levering op Santa Cruz geanalyseerd om de problemen te identificeren die bijdrage aan de huidige situatie met betrekking tot de beperkte en onregelmatige levering van water.

Resultaten laten 3 bronnen van waterlevering zien: (1) brak grondwater gedistribueerd door de gemeente, (2) gebottled-ontzilt grondwater geleverd door kleine private partijen en (3) brak grondwater geproduceerd door private partijen en verkocht door middel van water trucks. Uit dit onderzoek, de geschatte hoeveelheid geleverd water per persoon (inclusief alle 3 de bronnen) is ca. 370 lpcpd, dat erg hoog is vergeleken met de water levering in de hoofdstad Quito, waar het 210 lpcpd bedraagt.

Daarna is de water behoefte geschat, met inachtneming verschillende behoefte categorieën zoals gezinshuishoudens, hotels, restaurants en wasserettes. De water behoefte is onderzocht door middel van de verspreiding van ongeveer 400 vragenlijsten (240 aan gezinshuishoudens, 29 onder hotels, 30 onder restaurants en 16 onder wasserettes) in Puerto Ayora. De kwantificering van de waterbehoefte is gedaan elk van de drie eerder genoemde waterbronnen.

Resultaten laten zien dat de gemiddelde behoefte per persoon van de gemeentelijke levering is geschat op 163 lpcpd dat erg hoog is voor een eiland zonder vers-waterbronnen en met een onregelmatige levering van 3 uur per dag. Later zijn, om de specifieke gezins-behoefte te verifiëren, 18 watermeters geïnstalleerd in Puerto Ayora. Daarnaast zijn de resultaten geanalyseerd van ongeveer 300 door de gemeente eerder geïnstalleerde watermeters in 3 pilot zones en vergeleken met de resultaten van de 18 watermeters. Beide resultaten lieten een zeer hoog binnenlandse waterconsumptie zien. Het gemiddelde van de 18 watermeters registreerde een behoefte van 164 lpcd met een standaarddeviatie van +/- 94. De gemiddelde binnenlandse behoefte van de pilot zones was geschat voor resp. pilot zone 1, pilot zone 2 en pilot zone 3 182 lpcpd ±31, 195 lpcpd ±80 en 428 lpcpd ±70. In veel van de pilot zones kwamen uitschieters voor met een gemiddelde consumptie van 4500 lpcpd dat een excessief water verspilling binnen huishoudens en informele toeristische accommodaties suggereert.

Op basis van 4 verschillende groeiscenario's is er een voorspelling gegeven van stedelijke waterlevering en -behoefte voor de komende 30 jaar, Deze scenario's waren: (1) langzame groei- 1% voor lokale populatie en toerisme, (2) gemiddelde groei- 3% voor lokale populatie en 4% voor toerisme, (3) snelle groei- 5% voor lokale populatie en 7% voor toerisme (de huidige situatie) en (4) zeer snelle groei- 7% voor lokale populatie en 9% voor toerisme. Voorlopige resultaten lieten zien dat zonder interventie ('business as usual') de dekking van de waterbehoefte nauwelijks de 50% zal bereiken voor wat betreft het langzame groeiscenario en 10% voor het zeer snelle groeiscenario.

Vervolgens zijn 5 interventie strategieën voorgesteld en geëvalueerd als mogelijke opties om de huidige en toekomstige water schaarste op te lossen. De strategieën omvatten duurzame opties zoals opslag van regenwater, recyclen van brak water, reductie van lekkage, installatie van watermeters en reductie van de waterbehoefte per huishouden, alsmede het bouwen van een ontziltingsfabriek het eiland. Deze strategieën zijn geëvalueerd middels diverse KPI's zoals dekking water behoefte met waterlevering, kosten en energieverbruik in de komende 30 jaar. De resultaten lieten zien dat de interventiestrategie waarin het bouwen van een ontziltingsfabriek was opgenomen, de enige strategie is, waarbij er een 100% dekking van de waterbehoefte tot stand komt en waarbij de waterkwaliteit significant zal verbeteren aan het einde van de planningshorizon. Echter, deze strategie bleek eveneens de duurste en degene met het meeste energieverbruik. De interventiestrategie die in alle andere opties voorzag behalve het bouwen van een ontziltingsfabriek, zouden voldoen maar alleen in het langzame groeiscenario, dat erg onwaarschijnlijk is.

Daarna is een Multi-Criteria Beslissingsanalyse uitgevoerd, met als doel de meest duurzame interventiestrategie vast te stellen voor toekomstige waterleveringsproblemen in een gemiddeld groeiscenario. De strategieën zijn gecategoriseerd en geëvalueerd middels 4 criteria: milieu, technische, economische en sociale, en 4 groepen 'stakeholders' in ogenschouw nemend: lokale beleidsbepalers (gouvernementele autoriteiten), lokale experts (onderzoekers, academici en milieudeskundigen), binnenlandse eindgebruikers en hotels.

De voorgestelde strategieën zijn uiteindelijk gerangschikt op basis van de voorkeuren van de stakeholders, resulterend in de voorkeursstrategie op basis van geselecteerde criteria. De resultaten verschilde voor iedere stakeholder groep: voor lokale experts en lokale beleidsbepalers lag de voorkeur bij de strategie met het bouwen van de ontziltingsfabriek, terwijl voor de binnenlandse eindgebruikers en hotels die strategie als laatste de voorkeur had omdat deze groep een meer milieuvriendelijke oplossing prefereerde.

Vervolgens is een sensitiviteitsanalyse gedaan met als uitkomst dat de meest gevoelige criteria liggen op het gebied van sociale-, technische-, en milieuaspecten. Een kleine wijziging in de wegingsfactoren van deze criteria, dat het resultaat was van de feedback van de stakeholders, zou een significante wijziging in de ranking van de interventie strategieën tot gevolg hebben.

Ook is het waterleveringsnetwerk van Santa Cruz geëvalueerd met behulp van EPANET software, met als doel de huidige en toekomstige prestaties te onderzoeken in verschillende groeiscenario's gebruikmakend van 'Demand-Driven Analysis' (DDA) en 'Pressure-Driven

Analysis' (PDA) benaderingen. Elke benadering veronderstelde dat het huidige netwerk voldoet voor een 2- uurs levering in termen van kwantiteit en waterdruk binnen het netwerk. Op dezelfde manier zijn de opslagfaciliteiten van de huishoudens onderzocht, en is een methodologie ontwikkeld (gebruikmakend van de Emitter Coëfficiënt toepassing van de software) met als doel de overstroming van de daktanks de kwantificeren in verschillende scenario's van water consumptie, lekkage en tank. De resultaten onthullen significant verlies van water variërend van 5 tot 32% van de dagelijkse waterlevering, dat een significante hoeveelheid waterverlies is binnen de huishoudens.

De geschatte waterbehoefte, als ook de waterverspilling binnen de huishoudens is verrassend hoog voor een eiland waar er geen vers waterbronnen zijn en waterlevering van 3 uur per dag met tussenpozen. Hoogst waarschijnlijk is dit een direct gevolg van de vaste tarief structuur, dat voor de bewoners geen 'incentive' is om water te besparen. Toekomstige groei in het aantal toeristen moet door de overheid worden gereguleerd. Als het zeer snelle groeiscenario als overheidsdoel wordt gedefinieerd (jaarlijks 9%), zal, onvermijdelijk, een ontziltingsfabriek moeten worden geïnstalleerd. Dit zal kwaliteit- en kwantiteitsproblemen volledig oplossen, maar daarentegen een aantal negatieve milieu effecten veroorzaken, voornamelijk op het gebied van zout afvoer, chemische afbraak, energie consumptie en brandstof import. Voor de snelle en gemiddelde groeiscenario's zouden de voorgestelde houdbare oplossingen (inclusief regenwateropslag, brak waterrecycling, reductie van lekkage, watermeter installatie en reductie van de waterbehoefte per huishouden) gedeeltelijk afdoende zijn, vooropgesteld dat tenminste voor sommige activiteiten ook een kleinere ontziltingsfabriek gebouwd zal moeten worden. Alleen voor het langzame groeiscenario, zou de voorgestelde interventiestrategie die duurzame strategieën omvat, voldoende zijn, aannemende dat het overheidsdoel met betrekking tot toerisme zal moeten worden aangepast of zelfs teruggedraaid. Als de huidige groeitrends zich doorzetten, zal de overexploitatie van brak water van de basale aquifer waarschijnlijk het zoutgehalte van het geleverde water doen toenemen, waarmee de behoefte groeit voor het bouwen van een ontziltingsfabriek. Daarom moet de hoeveelheid verspild water op het eiland worden vastgesteld, moeten vaste watertarief structuren worden afgeschaft en moet de lokale populatie en het aantal toeristen dat het eiland bezoekt worden gereguleerd om dit fragiele en unieke ecosysteem te behouden. Echter, het bouwen van een ontziltingsfabriek, is misschien de meest uitvoerbare optie om alle water kwaliteitsproblemen het hoofd te bieden.

Resumen

Las Islas Galápagos, una de las provincias de la República del Ecuador, es un archipiélago volcánico de gran importancia ecológica, lo que le convierte en un lugar único en el mundo. Durante siglos, los recursos hídricos en estas islas han sido percibidos como escasos y, por ende, los problemas relacionados con la falta de agua, han sido reconocidos como una cuestión urgente. Como en muchas otras islas turísticas, los recursos hídricos se han visto seriamente amenazados por la elevada expansión de la industria turística. Además, los datos sobre el abastecimiento y la demanda de agua son escasos ya que la vasta investigación realizada en estas islas se centra principalmente en la conservación de especies endémicas, dejando a un lado el impacto de las actividades humanas sobre los este tipo de recursos.

La Isla de Santa Cruz, es el centro de turismo del Archipiélago de Galápagos. Esta isla tiene graves problemas relacionados con la cantidad y calidad del agua, ya que no existen recursos de agua dulce permanentes en esta isla, excepto el agua lluvia. Por lo tanto, el sistema municipal suministra agua salobre no tratada (800-1200 mg / l de cloruro) intermitentemente. Los problemas relacionados con el agua se sienten sobre todo en la ciudad de Puerto Ayora, donde se encuentran la mayoría de infraestructura turística, y donde además, la provisión media de agua municipal es de tres horas al día. En Puerto Ayora, no hay medidores de agua y la estructura tarifaria del agua es fija. El sistema de abastecimiento municipal no ha podido hacer frente al crecimiento actual del turismo y de población local (7% de crecimiento anual de turismo y 3,3% de crecimiento anual de población local). Las pérdidas de agua que van desde el 35% (esta tesis) hasta el 70% (estudios anteriores) en la isla, el mantenimiento deficiente y viejos sistemas de tuberías también se suman a la prestación de un servicio errático dentro de esta isla.

La investigación presentada en esta tesis se centró en la gestión del abastecimiento y demanda de agua en la isla de Santa Cruz. En primer lugar, se analizó el abastecimiento de agua con el fin de identificar los problemas que contribuyen a la situación actual en cuanto a la limitación e intermitencia del abastecimiento del recurso. Los resultados mostraron tres fuentes de abastecimiento: (1) agua salobre subterránea/ de grieta distribuida por el sistema municipal, (2) agua salobre embotellada-desalinizada producida por pequeñas empresas privadas y (3) agua salobre subterránea / de grieta privada y vendida por tranqueros. De este estudio, la cantidad estimada suministrada per cápita (incluyendo las tres fuentes) es de ± 370 lpcpd, lo que puede considerarse alto en comparación con el suministro en la ciudad capital de Quito, que es de 210 lpcpd.

Posteriormente, se estimó la demanda de agua considerando las diferentes categorías de demanda, tales como hogares domésticos, hoteles, restaurantes y lavanderías. La demanda de agua se evaluó a través de aproximadamente 400 encuestas (240 en hogares domésticos, 29 en hoteles, 30 en restaurantes y 16 en lavanderías), las cuales fueron distribuidas en todo el pueblo de Puerto Ayora. La cuantificación de la demanda de agua se realizó para cada una de las tres fuentes previamente identificadas. Los resultados mostraron que la demanda media per cápita del suministro municipal es ± 163 lpcpd, lo que puede considerarse alto para una isla sin recursos de agua dulce y suministro intermitente durante aproximadamente tres horas diarias. Posteriormente, para verificar la demanda doméstica, se instalaron 18 contadores de agua alrededor del pueblo de Puerto Ayora. Además, se analizaron las lecturas de aproximadamente 300 medidores de agua previamente instalados en tres zonas piloto establecidas por el municipio y se compararon con los resultados de los 18 medidores de agua. Ambos resultados mostraron un consumo de agua doméstico muy alto, el promedio obtenido de los 18 medidores de agua mostró una demanda promedio de 164 lpcpd ± 94 lpcpd de desviación estándar. La demanda interna promedio de las zonas piloto se estimó en 182 lpcpd ± 31, 195 lpcpd ± 80 y 428 lpcpd ± 70, para la zona piloto 1, la zona piloto 2 y la zona piloto 3, respectivamente. En muchas de las zonas piloto se identificaron valores atípicos con un consumo medio de hasta 4.500 lpcpd, lo que sugiere un desperdicio excesivo de agua en los hogares y / o alojamientos turísticos informales.

Un pronóstico de la oferta y demanda de agua urbana se llevó a cabo durante los próximos 30 años, para cuatro diferentes escenarios de crecimiento anual. Estos escenarios fueron: 1) crecimiento lento - 1% para la población local y turismo, 2) crecimiento moderado - 3% para la población local y 4% para el turismo, 3) crecimiento rápido - 5% para la población local y 7% Para el turismo (situación actual) y (4) crecimiento muy rápido -7% para la población local y 9% para el turismo. Los resultados preliminares mostraron que sin ninguna intervención (business as usual) la cobertura de la demanda de agua alcanzará apenas el 50% para el escenario lento y el 10% para el escenario muy rápido. En consecuencia, se propusieron y evaluaron cinco estrategias de intervención como opciones para resolver la actual y futura escasez de agua. Las estrategias incluyeron opciones sostenibles como la recolección de agua de lluvia, reutilización de aguas grises, reducción de fugas, instalación de medidores de agua y reducción de la demanda per cápita de agua, así como la instalación de una planta de desalinización de agua de mar en la isla. Estas estrategias fueron evaluadas utilizando varios Indicadores Clave de Desempeño (KPI), en términos de cobertura de la demanda de agua, costos y uso de energía durante los próximos 30 años. Los resultados mostraron que la estrategia de intervención, incluyendo la instalación de una desalinizadora, será la única estrategia que incrementará la cobertura de la demanda de agua al 100%, y mejorará significativamente la calidad del agua al final del periodo de estudio, pero será la más costosa y la mayor consumidora de energía. Sin embargo, la estrategia de intervención que incluía todas las opciones excepto la desalinización, será suficiente, pero sólo para el escenario de crecimiento lento, el cual muy poco probable que suceda.

Posteriormente, se realizó un Análisis de Decisión Multi-Criterio, con el objetivo de establecer en el futuro la estrategia de intervención más sostenible para los problemas de abastecimiento de agua, suponiendo un escenario de crecimiento moderado. Las estrategias fueron categorizadas y evaluadas bajo cuatro criterios: medioambiental, técnico, económico y social, y considerando cuatro grupos de actores locales: responsables locales (autoridades gubernamentales), expertos locales (investigadores, académicos y ecologistas), usuarios domésticos y hoteles. Las estrategias propuestas se clasificaron finalmente en función de las preferencias de los distintos actores locales, proporcionando la estrategia preferida teniendo en cuenta los criterios seleccionados. Los resultados difirieron para cada grupo de interesados: para los expertos locales y los responsables locales, la estrategia de intervención que incluye la instalación de una planta de desalinización, ocupó el primer lugar, mientras que para los usuarios domésticos y los hoteles esa estrategia quedó en último lugar y prefirieron opciones más amigables con el medio ambiente. Además, se realizó un análisis de sensibilidad, que mostró que los criterios más sensibles son los ambientales, técnicos y sociales, ya que un pequeño cambio en los valores de sus pesos, que fue el resultado de la retroalimentación de los actores locales, puede cambiar el ranking de las estrategias de intervención de manera significativa.

Además, se evaluó la red de abastecimiento de agua de Santa Cruz con el software EPANET, con el objetivo de evaluar el desempeño actual y futuro bajo varios escenarios de crecimiento, utilizando usando los enfoques de Análisis Derivado de Demanda (DDA) y Análisis Derivado de Presión (PDA). Cada enfoque sugiere que la red actual es suficiente para un suministro de 24 horas, en términos de cantidad de agua y presión disponible dentro de la red. Asimismo, se evaluaron las instalaciones de almacenamiento de los hogares, desarrollando una metodología (utilizando la característica Coeficiente Emisor del software) con el objetivo de cuantificar el desbordamiento de los tanques elevados usando varios escenarios de consumo de agua, fugas y tamaño de los tanques. Los resultados revelan pérdidas de agua significativas como resultado del desbordamiento de tanques, variando de 5 a 32% de la provisión diaria total, lo que representa una cantidad significativa de pérdida de agua en los hogares.

La demanda de agua estimada, así como el desperdicio de agua dentro de los hogares, es sorprendentemente alto para una isla donde no hay recursos de agua dulce y suministro intermitente durante tres horas al día. Lo más probable es que esto es una consecuencia directa de una estructura fija de tarifa de agua, la cual no provee incentivo alguno a la gente para ahorrar agua. Además, el crecimiento futuro en el número de turistas debe ser controlado por el gobierno. Si el escenario de crecimiento turístico muy rápido se define como un objetivo gubernamental (9% anual), inevitablemente, se debe instalar una planta de desalinización de agua de mar. Esto resolverá los problemas de cantidad y calidad de agua en su totalidad, pero muchos impactos ambientales negativos pueden ser generados, especialmente con respecto a la eliminación de salmuera, descargas de productos químicos, consumo de energía e importación de combustible. Para los escenarios de crecimiento rápido y moderado, las estrategias sostenibles propuestas (incluyendo la recolección de agua de lluvia, la reutilización de aguas grises, la reducción de fugas, la instalación de medidores de agua y la reducción de la demanda per cápita) serían parcialmente suficientes, sugiriendo que al menos para algunas actividades, una desalinizadora debería ser también instalada. Sólo para el escenario de crecimiento lento, la estrategia de intervención propuesta que comprende estrategias sostenibles sería suficiente, lo que sugiere que el objetivo gubernamental de turismo debe ser modificado o incluso frenado. Si continúan las actuales tendencias de crecimiento, la sobreexplotación del agua salobre procedente del acuífero basal puede aumentar la salinidad del agua suministrada, lo que impulsa la necesidad de una planta de desalinización. Por lo tanto, la cantidad de agua que se desperdicia en la isla debe ser seriamente analizada y solucionada, las tarifas fijas del agua deben ser abolidas y la población local y el número de turistas que visitan la isla deben ser controlados para preservar este ecosistema frágil y único. Sin embargo, una planta de desalinización de agua de mar puede ser la opción más viable para resolver los problemas de calidad del agua por completo.

"In a sense, each of us is an island. In another sense, however, we are all one. For though islands appear separate, and may even be situated at great distance from one another, they are only extrusions of the same planet, Earth."

—J. Donald Walters

1

GENERAL INTRODUCTION

1.1   Water resources on small islands

Over 100,000 islands are homed by planet Earth, which account for approximately 20% of global biodiversity (Richardson 2017). They are particular because of their uniqueness, which is further characterized by their size, shape and degree of isolation. Because of this, they have become ecologically and culturally important. Nevertheless, these characteristics also contribute to their fragility and vulnerability.

Islands around the world vary in size, geology, climate, hydrology, type of water resources, distance to the mainland, etc.; but they do have one issue in common: water stress problems and the consequent challenges that these have brought up. According to Bruce et al. (2008), islands are characterized mainly by the limitation of their water resources, which is a direct result of their insular condition, as well as their geological formation. Furthermore, among all islands, small islands are considered one of the most vulnerable human and natural systems, because of their size, limited natural resources, their remoteness, rapid population growth, and climate variability. These small islands, as well as small island states and small island developing states (SIDS) are located usually in the tropics and subtropics. In general, they consist of regions of the Pacific, Indian, and Atlantic Oceans, as well as the Caribbean and Mediterranean Seas (Nurse et al. 2001). Even though the different types of islands are not considered as a homogeneous group, they do share their vulnerability.

Most of small islands are rich in biodiversity, resulting in endemic flora and fauna. However, they tend to have relatively few natural resources. The resulting scarcity of water resources could be attributed to the climatic and physical conditions of each specific island. For instance, the amount of freshwater available in island communities depends on a delicate balance between consumption and climatic, hydro-geological, and physiographic factors (White and Falkland 2010). For many islands, the main problem is their semi-arid conditions, resulting in low precipitation levels. For others, such as volcanic islands, the lack of freshwater has been a persistent problem, depending on their availability (only) on their rainy seasons (d'Ozouville et al. 2008a). Nevertheless, even in islands with high precipitation rates, water resources are considered scarce due to the limited capacity of storage in the dry seasons, and their relative limited surface area (Hophmayer-Tockich and Kadiman 2006). Consequently, a serious and restrictive factor relies upon the availability of freshwater resources in the majority of inhabited small islands (Kechagias and Katsifarakis 2004).

Furthermore, many small islands worldwide are completely arid and their water resources are further limited. These islands are also experiencing excessive population growth and densities, which increase water demand and potential pollution of their water resources (Tsiourtis 2002). According to the Asian Development Bank, the size of islands limits their availability of natural resources, including water. Because of this water scarcity, the local economic growth is in many occasions restrained, which is mostly the tourist industry. Tourism is the most important motor of the economy for some islands (e.g. Barbados, Bahamas, Malta, Galápagos, etc.), and for others is the second most important economy (e.g. Maldives, Western Samoa, etc.) (Ghina 2003). Nonetheless, the assurance of their development and success of economic activities depends greatly on the preservation of their scarce natural resources. The main characteristics of small islands include this dependence on tourism for their economic prosperity, as well as their limited infrastructure, low financial means, lack of skilled human resources and high population growth (Ghina 2003).

The freshwater availability is mainly attributed to the geology and formation of the different islands. In case of volcanic islands, the lack of freshwater is significant. Their main source is brackish groundwater, which is the mixture of intruded seawater and rain in the basal aquifers (d'Ozouville *et al.* 2008a). Furthermore, their water availability may also vary depending on the levels of precipitation, making some regions more water scarce than others. In other type of islands, the main source is groundwater, such as in the Caribbean Islands like the Bahamas, Barbados, Jamaica and St. Kitts. On the other hand, other islands have surface water as their main supply, such as St. Lucia and Trinidad & Tobago (Ekwue 2010). Several Mediterranean islands, such as the Greek Aegean, as well as Malta, use desalinated seawater as their main source (Kondili *et al.* 2010). In extreme situations, freshwater is imported on tanker ships from the mainland like in the Bahamas, Antigua, Mallorca and some Greek Islands (UNESCO 2009). According to White and Falkland (2010) there are about 1,000 populated small islands in the Pacific Ocean, which have their main source of freshwater from groundwater, but it is limited and compromised also by pollution. In addition, the intrusion of salty water into groundwater resources is a common and major problem (Khaka 1998). This intrusion affects not only the quality of the water but also the quantity, since the transition-gradual mixing zone of fresh and salt water is extensive (Gingerich and Oki 2000). Besides, these scarce water resources could be further threatened by climate change, climate variability and consequent rising sea levels (Bruce *et al.* 2008).

According to Donta and Lange (2008), small island states are also restricted by the lack of organizational expertise, deficient financial infrastructure, few incentives for water harvesting, old and unreliable water infrastructures leading to high leakage levels, and absence of effective water pricing and cost recovery systems. Due to all these issues, water supply alternatives in small islands need to be explored (Gikas and Tchobanoglous 2009). The literature points to several technical options to increase availability of water, especially addressing desalination (Kondili *et al.* 2010, Castillo-Martinez *et al.* 2014). Also, several environmental measures have been suggested by Hof and Schmitt (2011) to create awareness in order to reduce urban water consumption, as in the case of the Balearic Islands, as means to reduce the need of more water supply. However, each island is particular and therefore the set of solutions may vary according to each study area.

1.2 Tourism influence on small islands

Tourism in many tropical islands has increased dramatically over the last decades. It is the most dominant economic sector in several island states and other small islands, such as in the Caribbean (Charara *et al.* 2010). Even though tourism is a major source of income and employment for many islands (Briguglio 2008), it has been amongst the main causes of environmental degradation which exerts a significant pressure on natural resources, such as water. According to Mangion (2013), tourism is the main cause of extreme demand for diverse natural resources, inflicting environmental threats and high infrastructural costs, which unfortunately are often not taken into account. Even though tourist visitors contribute significantly to their economic growth, they often require services and facilities that produce an unsustainable balance between infrastructure and natural resources. Therefore, a massive construction due to tourism has been identified, as well as an insufficient control of urban planning, such as the case of many of the Mediterranean Islands (Essex *et al.* 2004).

1.2.1 Water use related to tourism

Despite the fact that islands have limited natural and water resources, the expansion of tourism to these type of destinations over the last 40 years has been overwhelming (Essex *et al.* 2004). As a result, tourism growth has caused in many island states, deficiencies in the provision of water supply and sewerage systems (Bramwell 2003, Marques *et al.* 2013). In addition, tourism increases overall per capita water consumption, concentrating it in time (often in the dry and high season). The proportional increase of water demand is also amplified by the growth of local population, which needs to provide facilities and services to satisfy tourist visitors. For instance, in some areas of the Mediterranean Islands, the ratio of local population to tourists may change over the year, reaching in some cases more than one to six. Specifically, in the Balearic Islands, water use during peak months of tourism (e.g. in July 1999) was equal to 20% of the water use of the entire local population for the whole year (Gössling *et al.* 2015). According to Mangion (2013), a tourist in Malta consumes on average three times more water than a local resident, creating a challenge for water supply utilities to be able to comply with these elevated rates (Briguglio 1995).

Water consumption in the tourism sector is varied. It comprises of several activities such as: bathing and showering, toilet use, golf courts, landscaping, spas, wellness areas, swimming pools, as well as food and fuel production. In addition to these uses, hotel water consumption includes recreational activities such as sailing, diving and fishing (Gössling *et al.* 2012). In consequence, the proportion of water consumption by the tourist sector can be as high as 40% (e.g. in Mauritius) over the resident average water use. This number tends to be higher in areas where water is scarce and the number of tourists is even higher.

Table 1.1-Water use categories and estimated use per tourist per day

Water use category	Description	L/tourist/day
Direct use	Accommodation (shower, bath, toilet, spas, pools, landscapes, sport and health centers, laundry, restaurants)	84-2000
	Activities (e.g. golf, skiing)	10-30
Indirect use	Fossil fuels (water use for energy consumption and/or production)	750 (per 1000 km by air/car)
	Biofuels (water use for biofuel consumption and/or production)	2500 (per L of bio fuel)
	Food (water use for food consumption and/or production)	2000-5000
Total per tourist per day		**2000-7500**

Source: Gössling *et al.* (2012)

Tourism water use tends to increase with hotel standards, increased water-intense activities, as well as with the growth rates of tourist visitors. Even though the water consumption in hotels has been already recognized as high, it can result in even greater quantities in luxurious hotels. This type of hotels use significant amounts of water for recreational activities such as swimming pools, golf courts, extensive gardens, etc. Gössling (2001) reported that the variation on water consumption is directly related to the hotel category. For instance, in Zanzibar, the average water demand in guesthouses is 248 L/tourist/day, which is significantly lower than that in five-star hotels (931 L/tourist/day). According to Deng and Burnett (2002), the average water use in five-star hotels is approximately 5 m³/m² of hotel area, while in four-star hotels and three-stars hotels are 4 and 3 m³/m², respectively. According to this study, the main water consuming activities are: (1) garden irrigation, (2) swimming pools, (3) spas and wellness facilities, (4) golf courses, (5) cooling towers, (6) guest rooms and (7) kitchens. Furthermore, indirect water use related to tourism is often not taken into account, but this accounts for great quantities of water as well. Table 1.1 shows the average per capita consumption related to direct and indirect water use in the tourist sector.

In conclusion, the increase in water demand on islands is mainly attributed to the tourist sector. For example, in the Mediterranean Islands an average tourist consumes between 440-880 litres per capita per day (lpcpd) (Gössling et al. 2012), while in Jamaica, Barbados, St. Lucia and the Philippines the reported specific demands are 992, 756, 662 and 1499 lpcpd, respectively (Charara et al. 2010). The locals, on the other hand, consume on average as follows: in Mediterranean Islands 200 lpcpd, in Jamaica 160 lpcpd, in Barbados 200 lpcpd, etc. Because of the population growth and consequent per capita demand, water scarcity and stress is experienced, meaning that (ground) water resources are being overexploited. Overexploitation of aquifers are of great concern, and have been occurring in Malta, Tenerife, among many other islands (Axiak et al. 2002, Guilabert Antón 2012).

1.2.2 Water-related problems in islands and their relevant solutions

The tourist industry presents a significant problem for water supply utilities on islands, since tourist visitors often arrive during the dry season, when rainfall levels are low and water availability is at the minimum level. On islands where tourists are high water consumers, the scarcity problem may be worsened by climate change. According to (Gössling et al. 2012) water demand is likely to increase in the future due to increase in tourist arrivals, higher average tourist water consumption, as well as more water intense activities.

In addition, groundwater resources have become extremely vulnerable to pollution, as a result of poor sewerage and wastewater treatment infrastructures on islands with highly permeable soils (Hophmayer-Tockich and Kadiman 2006). Many coastal aquifers are also exposed to the increase in salinity levels as a result of sea level rise and the overexploitation of aquifers. Because of this, desalination has been considered as a viable option in order to preserve the water resources. However, with this option, energy consumption might increase, as well as environmental effects. Moreover, many areas may not be connected to the electric grid, and therefore the process becomes highly dependable on fuel importation to run generators. Also, this option means a high investment, which many islands cannot afford. Nonetheless, combined grid-renewable energy desalination plants can produce potable water with lower emissions (Gössling *et al.* 2012). For many islands, desalination is considered as the last resort solution, and other more economical and less expensive water supply strategies are considered first (Hophmayer-Tockich and Kadiman 2006).

Other strategies include demand side measures which aim to reduce the water demand. These include Non-Revenue Water (NRW) reduction, water metering, adequate water tariff structures based on water usage, the use of water saving devices, and wastewater recycling for non-potable use (Sharma 2014). Most of these measures are economical and can reduce notably the water use. Moreover, investment in sustainable technologies, as well as in water conservation and management, are key strategies that may avoid to resort to desalination and other strategies which are not considered so environment-friendly. However, in order to achieve this, strong policies are required, including the use of appropriate economic incentives and adequate water pricing to create environmental awareness and encourage water conservation (Gössling *et al.* 2012). Also, public information, as well as education are necessary tools for the successful implementation of water conservation programmes and activities.

1.3 Water supply and demand issues on small islands

Water demand is often considered to be equal with the water consumption term, but demand depends on many interlinked factors and features. Even though, theoretically, both terms add up to an equal amount, in practice, there are several differences between these two terms. Trifunovic (2008) identified the water consumption to be the quantity of water that is directly utilized by the various categories of consumers. On the other hand, water demand depends on several factors such as: climatic conditions, size and type of settlement, different standards of living, water supply quantities, pressure along the distribution system, supply regime (intermittent or continuous), water costs and tariff structure, metering, quality of water, extent of industrial and commercial activities, existing sanitation/sewerage, availability of private water supply, environmental issues, etc. (Sharma 2014).

According to Sharma (2014), urban water demand consists of domestic, commercial and industrial, public use, and miscellaneous (system cleaning, losses, fire demand) components. The domestic component reflects the populations' needs and consumption patterns, while the public and miscellaneous use determines the level of public awareness and system maintenance. However, commercial and industrial demand varies seasonally and annually.

Water demand estimations are usually based on measurements on the consumer side. These measurements, often, do not include leakages along the distribution system, since water meters are installed at service connections at the entrance of the households. This can often lead to the inaccurate demand estimation. For that reason, the measurements at various supply points within the system should be established with the purpose of taking more accurate water demand figures.

A proper analysis, including quantification and projection of water demand, is fundamental for an appropriate water supply infrastructure. If this is not done properly, it can result in serious issues, as well as difficulties and calamities in the water supply system operation. Those difficulties are often reflected in insufficient or overestimated capacities and in incorrect dimensions of the components of the supply system, which may provide the consumers with unreliable and inadequate provision of water.

It is imperative to understand also the temporal variations in the system, in order to determine the demand more accurately over time. Variations are commonly expressed as the peak factors (ratio of the present demand over the average total demand in a certain period of time) (Trifunovic 2008). Daily demand patterns depend on several factors such as: distribution area, percentage of the various demand categories, level of leakages, character of the supply (continuous/intermittent), etc.

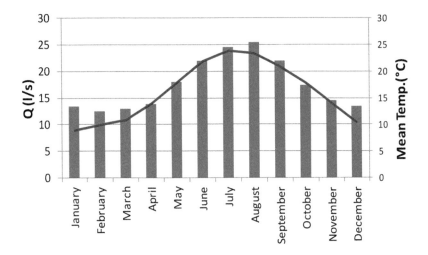

Source: Aquaprojekt and Hidroekspert (2001)

Figure 1.1- Annually pumped water in the island of Vis-Croatia

During the last decades, tourist islands have been dealing with enormous challenges regarding the water supply of the increasing demands. The different tourist activities within islands create a water demand that clearly exceeds the amount that can be supplied by local municipalities or water supply utilities (Reyes *et al.* 2016). Because of this, water suppliers on islands are facing challenging constraints in order to sustain water and wastewater provisions (Briguglio 2008). For instance, as shown in Figure 1.1, in the island of Vis, the average monthly pumped water from the source, indicated on the left Y-axis, increases significantly during the high season (May-September), and this coincides with the tourist seasonal variation in this island, indicated by the highest temperatures indicated in the right Y-axis. Therefore, as observed, the water pumping in the month of August is about 50% higher than the pumping in the month of February.

Because of these annual peaks of water demand in many islands, the water stress has been overcome through desalination or water transportation from other islands or the mainland. However, most of these options are considered rather expensive. For example, in some arid islands, such as the Greek Islands, the dependence is only on non-conventional water resources, such as desalination or water transportation (Liu *et al.* 2012). These options are often considered as non-sustainable and have low cost-benefits. Consequently and not surprisingly, the water demand has increased since there is more availability of water, increasing the environmental threats (Kondili *et al.* 2010).

1.4 Water (demand) management on islands

Urban water management has two major approaches. The traditional one is the supply-driven management, and the second one is the water demand management approach. The supply driven management mitigates the increasing demand by increasing the supply of water, looking for additional sources and constructing new capacities for the supply system. Yet, this approach has led to the over-utilization, over-capitalization, pollution increase and many other environmental issues, causing it to be obsolete and unsustainable.

As a result, the Water Demand Management (WDM) approach is considered as a more sustainable measure. According to Sharma (2014), it places the demand itself as the focus of the problem and "refers to any socially beneficial action that reduces or modifies average or peak water withdrawals or consumption consistently with protection or enhancement of water quality." This suggests that by acting on the demand and consumption side, through the introduction of several measures, water use can be significantly reduced and therefore, the increase of production capacity can be avoided.

Urban water demand management (UWDM) measures, according to Sharma and Vairavamoorthy (2009) can be divided into three main categories:

1-Structural and technical measures – reduction of the non-revenue water (water losses), using water saving devices, water meter management, etc.

2-Economical measures – different water tariffs and prices, various incentives, subsidies, penalties and fines to promote water conservation.

3-Socio-political measures – comprise mainly the legal framework that defines the regulations in favour of promoting the demand management measures and water conservation, public education and awareness building.

UWDM measures require an integral and multidisciplinary approach, which can be favourable to the supply system in many ways. These measures can balance in a sustainable way the water supply and demand on the long-term. However, in order to implement successfully these measures, the supply system needs to be known in detail, as well as socio-political and economical features of the society in which they will be applied. The society of small touristic communities has its own particularities, which determine the range, scope and extent of the application of water demand measures. According to Briguglio (2008), this should be investigated in much detail in order to make a positive and long-term impact.

Water management on islands is unique and complex as they are constrained by their size, isolation from the mainland, fragility and limited financial resources (Hophmayer-Tockich and Kadiman 2006). However, WDM and water conservation practices within islands are concepts that have been already approached by many Spanish islands. For instance, they have addressed mainly hydraulic problems by a group of activities that reduced the demand of water, developing efficient use and avoiding the reduction and depletion of water resources. In many cases, such as in Tenerife, water supply and water management has been carried out by private concessions, having a private provider in charge of the treatment, distribution and supply, as well as agreed pricing policies; thus, ensuring an optimal service and optimal quality. This privatization of water supply utilities provides a better service and improved population perception (Guilabert Antón 2012).

In many islands there is a need for the development of new guidelines, involving the design and control of (intermittent) water distribution systems. Also, network analyses are necessary to develop optimal and reliable designs, in order to reach equity of the water supply, as well as people driven levels of service (PDLS) (Vairavamoorthy *et al.* 2008). Therefore, several options for water supply have been analysed in the specific case of islands, where many alternatives have been successful for meeting the demand of water. Nevertheless, because of increasing tourism rates, the current and alternative sources may not be enough in the long-term. For these reasons, optimization of the management of local water resources is necessary, which includes the implementation of water demand-side management strategies, allowing the balance between supply and demand. Special emphasis must be placed on efficient and sustainable allocation and use of existing water supply, as well as to the design of adaptation strategies for the future.

1.5   General perspective of Santa Cruz Island

Freshwater resources are scarce in the Galápagos Islands (Ecuador), and the very few found, vary significantly based on the amount of precipitation contributing to surface run-off and basal aquifers. Thus, the amount of precipitation is of extreme importance since it is highly variable due to the climactic condition of the Galápagos. Despite of this limitation, water demand keeps increasing without any control. Santa Cruz Island, the biggest tourism centre in this archipelago has no significant source of freshwater, in the form of permanent surface or groundwater. Currently, water supplied is brackish-groundwater.

This group of islands are famous because of their natural beauty and uniqueness. Due to their environmental attractiveness, which include endemic species, they have been seriously threatened by the exponential increase in tourism and local population. Local and tourist premises are developing unsustainably with annual rates between 8 and 11%, resulting in an increasing number of immigrants over the last 20 years (INEC, 2010). Furthermore, annual local population growth in Galápagos varies between 3.32% and 5.87% from 1950 until 2010. Consequently, the demand for public services has also increased, resulting in lack of efficiency in the water supply service.

The main source of supply in Santa Cruz, which is extracted from the basal aquifer, is non-potable due to high concentration of chlorides (800-1200 mg/L), and is further distributed without any prior treatment (d'Ozouville, 2007). In addition, the water is also contaminated with *Escherichia Coli*, due to the lack of sewerage system (d'Ozouville *et al.* 2008b). Wastewater is mainly disposed into septic tanks, which are installed individually by each household. It has been identified that most of these septic tanks are not constructed technically, resulting in infiltration to the water sources (Liu and d'Ozouville 2013).

Furthermore, the water supply is intermittent and rationed, with an average supply of 2-3 hours per day (Reyes *et al.* 2015). Also, there is an excessive water loss in Puerto Ayora, which is caused by aging pipes and lack of maintenance in the distribution system. This problem has further worsened by the lack of water meters in the town of Puerto Ayora, which results in excessive water wastage within the households. This is also a consequence of the lack of water saving practices or specific policies promoting water conservation in this fragile ecosystem. In addition, there are also financial constraints that contribute to the difficulties faced by the municipality to improve the water service. On top of this, the lack of communication among different entities, and unclear distribution of responsibilities within the institutions regarding water management contribute to the situation (Reyes *et al.* 2016). Lastly, there is a significant absence of data and information, causing difficulties in the assessment of the current situation and the future planning for the improvement of the water scarcity.

1.6 Motivation of the study

The Galápagos Islands, the source of inspiration for Charles Darwin's '*The Origin of Species*' due to its "mystery of mysteries" condition, where surprisingly and amazingly, species adapted to a complete different environment. These characteristics created an exceptional place in the planet, which, unfortunately, is undergoing extreme stress in terms of their water resources. Because of the extraordinariness and the environmental significance of this place, which is considered a living and natural laboratory, the number of tourists has increased dramatically and consequently the local population (both legal and illegal). Nowadays, the islands, especially Santa Cruz (the principal tourism centre of the Archipelago), are confronted with severe lack of proper water supply reflected in intermittent and non-drinkable i.e. low quality and contaminated water, and the lack of proper sanitation. This is a direct consequence of (i) aged and unreliable water distribution networks and (ii) lack of proper management of the water supply system.

The lack of information and data on water supply, as well as on water consumption on the islands, has been identified in many studies. Santa Cruz Island is in urgent need of water management solutions, which require assessing the magnitude of the problem accurately. Specifically to current and persisting water supply and sanitation problems, there are no appropriate and systematic solutions developed yet. The available literature that could serve as a basis for this research is insufficient both in quality and quantity. Not surprisingly, most of the research references point to the issues of conservation of wildlife, reiterating the need for research in the direction of urban water demand growth and its impact and consequences on the environment. Therefore, an optimal solution is needed regarding human interactions, which will benefit the population in terms of assuring water in the future, and also preserving the fragile ecosystem. A sustainable balance between these two needs to be developed to assure water supply in the future.

Although some scattered studies related to the water resources have been done in the past, a complete assessment of the current situation, including all important aspects of water supply in Santa Cruz, is yet to be conducted. Research has been carried out regarding the nature of the water resources, as well as the supply issues and bacteriological contamination (d'Ozouville and Merlen 2007, d'Ozouville *et al.* 2008a, d'Ozouville *et al.* 2008b, Pryet 2011, Violette *et al.* 2014). Another research by Guyot-Tephiane (2012) addressed the perspectives, usage and management of water in Galápagos.

Yet, the implementation of integrated management measures and solutions are still absent. The Galápagos Islands are in urgent need of tangible water management solutions, which require reliable data and an integrated approach to the water supply and demand, in order to assess the magnitude of the problem accurately. The existing initiatives include those carried out by Water Management International, which are monitoring water losses and the (pilot) installation of water meters. Though there is more literature on the issue of water resources in the Galápagos Islands than before, there is a lack of integrated and systematic cross-checking of data. This is a direct consequence of local institutions not conducting any follow up of previous research, studies and consultancy. In such an environmentally fragile ecosystem, researchers tend to conduct studies towards conservation and environment, becoming less aware of the link between human impacts on water and the direct relation to environment degradation. The environment is affected by overexploitation of sources for water supply and discharge back of untreated wastewater, therefore this research points the relevance of achieving good quality information on the water balance between supply and demand, so that the measures can be taken with greater confidence.

1.7   Outline of the thesis

This thesis is structured in nine chapters that include the introduction (Chapter 1) and conclusions & recommendations (Chapter 9). The chapters which are the body of the thesis (Chapters 2 to 8) have the following order:

Chapter 2 describes the study area, providing information including the location, physical characteristics, institutional issues and water resources management.

Chapter 3 analyses the current water supply system, estimating the amount of water supplied from the different types of sources. Also, it explains the dynamic and processes of the water distribution system and provides an estimate of NRW.

Chapter 4 presents the results of a survey carried out in Santa Cruz Island, with the aim of quantifying water demand from different sources and categories (in the absence of water metering within premises), and understanding consumers' behaviours and attitudes regarding the preservation of this resource.

Chapter 5 evaluates the domestic water demand in detail, calculates ranges of consumption and assess water appliances' use, and compares the results with schedules and zones of water distribution established by the municipality. Also, it draws water demand patterns and compares water supplied and water demand in 2013, 2014 and 2015.

Chapter 6 forecasts water demand and supply using the WaterMet^2 model, under four different population growth scenarios. It elaborates on five different intervention strategies, presented as means to solve the water deficiency in a planning horizon of 30 years. This chapter presents in detail water demand, leakages, and percentage of water demand coverage with supply, energy consumption and costs.

Chapter 7 evaluates the intervention strategies and their results presented in the previous chapters with a Multi-Criteria Decision Analysis (MCDA), with the aim of assessing the best solution considering the environmental, technical, economic and social criteria, developing specific indicators. Also, it presents the results of an uncertainty analysis as well as a sensitive analysis conducted in order to analyse the impact when changing the indicator values or the weights of the criteria on the final results.

Chapter 8 models the water distribution network of the main town of Santa Cruz, addressing the intermittent condition with Pressure-Driven Analysis (PDA) as means to estimate water demand more accurately based in the network's pressure. It also develops a methodology to estimate the amount of overflow of roof tanks, which are prevalent in this town, and evaluates whether they are indeed helpful or contribute further to the intermittent condition. Finally, it assesses the condition of the distribution network with future growth scenarios.

Chapter 9 synthesizes the main findings of the study and proposes recommendations for future research.

Some detailed information is provided in the Annexes section, corresponding to each chapter, as well as a short biography and the list of publications.

"The archipelago is a little world within itself, or rather a satellite attached to America, hence it has derived a few stray colonists, and has received the general character of its indigenous population."

- Charles Darwin

2

CASE STUDY DESCRIPTION

This chapter describes the Galápagos Islands in detail, as well as the creation of the different sources of water for the various islands. Then, it explains the problems of water scarcity in this Archipelago, and how this fact has been addressed so far. Furthermore, it elaborates on the water supply issues on the island of Santa Cruz, the hub of tourism of this group of islands. It introduces to tourism and local population exponential growth and the impact it has had on natural resources, especially on water. Also, summarizes research done so far on water supply and water demand in the previous years, and reveals the lack of data on this topic.

2.1 General information on the Galápagos Islands

The Galápagos Islands make up an archipelago of volcanic islands of prominent global ecological importance. It is the last well conserved tropical archipelago in the world and it is home to unique ecosystems. It was recognized as a UNESCO World Heritage Site in 1978 because at least a third of all their existing native species are endemic, meaning that they are only found within the islands, and nowhere else in the world (Watkins 2006).

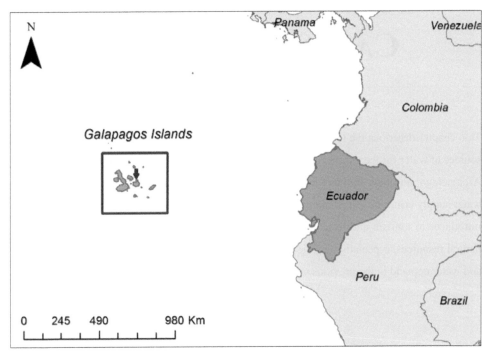

Source: (Reyes et al. 2016)

Figure 2.1-Map of Ecuador and the Galápagos Islands

They are located 600 miles off the coast of Ecuador, making up a province of this country (Figure 2.1), as indicated with the red box (the red arrow points to Santa Cruz Island, the case study area). Frequently, it is referred to as a natural laboratory due to the uniqueness of this ecosystem and the biodiversity encountered here (González *et al.* 2008). These enchanted islands have often been called or denominated as 'dry'. Limited water resources on the islands have been witnessed since their discovery by whalers and buccaneers, and more recently, through several investigations and studies. These limitations originate from the volcanic nature of the archipelago, and also because of their climatic variability, which makes the availability of resources scarce during some months of the year and abundant in others, and very variable from one year to the next.

There are 13 main islands, three smaller islands, and 107 rocks and islets in Galápagos (CDF 2016) as shown in Figure 2.2. This group of islands has an area of 8,010 km², of which 3.3% is available for human activities and the remaining 96.7% is under the jurisdiction of the Galápagos National Park and is reserved for the preservation of natural ecosystems of the islands. Owing to its unique flora and fauna, the islands have attracted increasing number of tourists in recent years, now reaching nearly 225,000 per year, creating a stress on all natural resources, especially on water, as well as on the fragile environment. Five are inhabited islands, each with their own water related problems and different sources of freshwater. San Cristóbal is the only island that has a constant freshwater source in the form of surface streams, a unique characteristic that differentiates it from the rest.

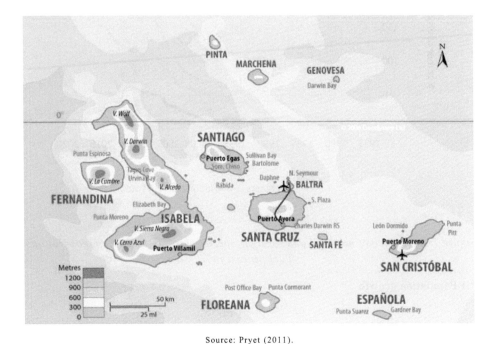

Source: Pryet (2011).

Figure 2.2 Geographic map of the Galápagos Islands with altitudes

The other inhabited islands have to deal with significant lack of freshwater, both surface and subsurface, and its consequences and impacts on local population. For example, Floreana depends on very small springs, and Santa Cruz and Isabela depend on the extraction from brackish basal aquifers or other sources such as seawater, which is also the case of the island of Baltra (d'Ozouville and Merlen 2007, Guyot-Tephiane 2012).

The quality, as well as the quantity of freshwater has been an ever-present problem in Galápagos (d'Ozouville and Merlen 2007). Research into the hydrological functioning of the islands was therefore initiated in 2003 by a team at the University of Paris (d'Ozouville *et al.* 2008b). This research links the geological-and biological sciences, relying on knowledge from one and providing information to the other about the physical conditions of ecosystems. Furthermore, these studies argue that water resources and their dynamics are an essential link between the natural systems and the human ones.

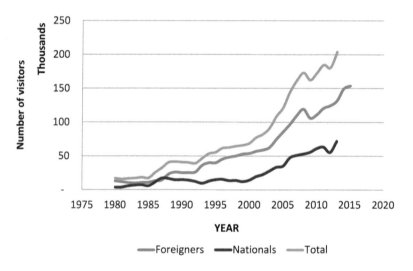

Source: Direccion del Parque Nacional Galapagos (2014)

Figure 2.3-Increase of national and foreign tourists entering the Galápagos National Park since 1980

2.1.1 Population growth

The recognition of the ecological uniqueness and conservation of Galápagos has enhanced the tourist activities significantly, becoming undoubtedly one of the biggest and fastest growing businesses within the islands. Moreover, the increased number of tourists has contributed also to a rapid local population growth, generating significant economic impact (Ortiz 2006).

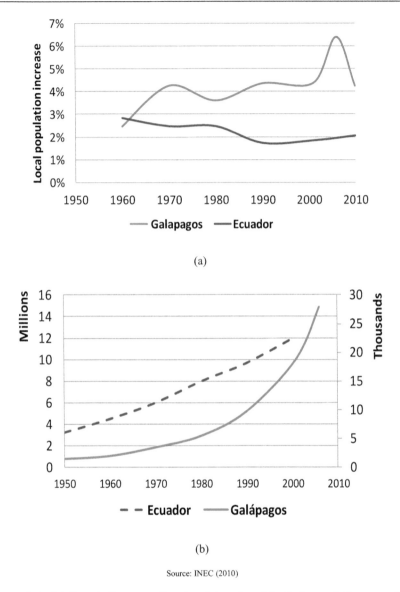

(a)

(b)

Source: INEC (2010)

Figure 2.4- (a) Population growth rates (a) and residential population growth in Ecuador and Galápagos Province

According to the Galápagos National Park records, in 2015 the islands welcomed 224,125 visitors compared to only 17,445 visitors received in 1980 (Figure 2.3). According to INEC (2010), there has been an exponential increase in the number of visitors and in the local population (Direccion del Parque Nacional Galapagos 2014).

In the 1970s, the 'floating hotel' model was the main form of tourism, where visitors stayed on ships and only brief visits were allowed to the different sites on the islands. Nevertheless, most of these vessels used to get (potable) water from Puerto Ayora. During the 1990s and the 2000s the tourism model changed to land-based tourism, which contributed to the upgrading of infrastructure and services in order to host more guests. Nowadays, most of the ships visiting the islands have their own desalination plants on board and put no more additional demand on scarce water resources.

The annual population growth rate in Galápagos was 4.9% from 1974 to 1982 and 5.9% from 1982 to 2001. From 1990 to 2001 the growth rate in the mainland was 2.1%, less than a half of that in the Galápagos province. In 1998, the Special Law of Galápagos was created with the aim of controlling the immigration to these islands. Because of this control and more strict regulations, the rate of growth reduced between 2000 and 2010 (INEC, 2010). Even though, the permanent population in Galápagos is small (~ 25,124 in 2010), it is considered dynamic and it continues to grow.

The current annual population growth rate is 3.32%, with an average density of 80 people per km^2. In comparison, the annual growth rate in Ecuador mainland is 1.95%, with an average density of 56 people per km^2 (Mena *et al.* 2013). Figure 2.4 shows population growth trends and residential population of the Galápagos province compared to Ecuador mainland.

Table 2.1 shows the population structure of the inhabited islands, divided into the main urban settlements on each island, according to the municipal records.

2.1.2 Tourism development

The Galápagos Islands have become a famous tourist destination, mainly because of their unique landscapes and species. However, the human presence in the archipelago has eliminated the main and special characteristic of isolation, altering the native ecosystem in a significant way. Due to the exponential increase of local population and number of tourists, which are the main causes of this alteration, saturation levels are getting reached. According to Watkins (2006), if tourism continues to grow at the same pace of the last two decades, it cannot be ensured that every visitor of Galápagos will experience environmental tourism with high quality standards.

Table 2.1- Permanent population of the main settlements in Galápagos in 2010

Municipality	Towns	No. of inhabitants	Percentage of total (%)
San Cristobal	Puerto Baquerizo Moreno	6,672	26.6%
	El Progreso	658	2.6
	Isla Santa Maria (Floreana)	145	0.6
	Total	**7,475**	**29.8**
Isabela	Puerto Vilamil	2,092	8.3
	Tuomas de Berlanga	164	0.7
	Total	**2,256**	**9.0**
Santa Cruz	Puerto Ayora	11,974	47.7
	Bellavista	2,425	9.7
	Santa Rosa	994	4.0
	Total	**15,393**	**61.3**
Galápagos	**Total**	**25,124**	**100.0**

Source: (Mena *et al.* 2013)

In order to ensure the protection of these ecosystems, as well as the welfare of the local residents, a sustainable balance of natural resources consumption by inhabitants and tourists, must be reached. However, the social, administrative and organizational authorities in the Galápagos Islands have not been able to cope with the current trends of expansion of tourism. As a consequence, this has created failures in the protection of this place and has expanded the negative impacts over the ecosystems. For instance, in June 2007, the Ecuadorian government declared the Galápagos at emergency and the UNESCO declared them "at risk", due to the high threats caused by the invasion of non-endemic species, exploitation of natural resources and exponential population growth associated with uncontrolled immigration from the mainland (Epler 2007).

In 1956 began the idea of the archipelago as a destination for environmental and elite tourism, which in turn promoted and supported conservation. For this purpose, the Charles Darwin Foundation and the Galápagos National Park were created. Later, in the late 1960s, a former military airport on Baltra Island was restored, initiating the operation of two flights per week. This enabled people to travel more easily between the islands and the mainland. However, the cruise-based tourism was the predominating type around the islands, where passengers stayed mostly on their cruise ships or boats (Epler 2007). This model was developed with the aim of promoting an environment-friendly form of tourism since it has less ecological footprint, requiring less infrastructure compared to the land-based tourism. In addition, land-based tourism was scarce since land infrastructure and services were not yet available in these islands, but until few decades later (only in 1998 was electricity available).

Table 2.2- Economic activities in Galápagos and Ecuador

Economic Activity	Galápagos (%)	Ecuador (%)
Wholesale and retail trading	12.8	18.4
Public administration and defense	10.7	4.1
Housing/lodging and alimentation	9.5	3.8
Agriculture, livestock, forestry and fishing	9.0	21.8
Construction	7.5	6.5
Storage and transport	7.0	5.2
Administrative services and support	7.0	2.7
Teaching/Education	5.6	5.1
Manufacture industry	5.1	10.2
Other	25.8	22.1
Total	**100.0**	**100.0**

Source: (Mena *et al.* 2013)

The major expansion of tourism occurred between the mid-1970s and 1980s. Figure 2.4(b) shows the evolution of tourism over the last five decades. This was caused by the declaration of the Galápagos as a World Heritage Site by UNESCO in 1978, which attracted the interest of mainly nature lovers (Epler 2007). Therefore, in order to meet the growing rate of tourist visitors, the number of cruise ships, boats and flights increased almost tenfold. The critical economic conditions in Ecuador from 1980 to 1985, stabilized somehow tourist numbers. But later in the late 1980s, a cheap land-based tourism began to flourish, attracting tourists from different socio-economic levels. In the 1990s, some concerns were pointed out about the exponential population growth caused by the expansion of land-based tourism, which led to an unsuccessful attempt to return to the cruise-based tourism only (Giampietro *et al.* 2012). Finally, in 1998, the Special Law for Galápagos was created with the aim of stopping the high migration from the mainland, as well as foreigners staying illegally. Also, this new law caused a higher percentage of tourism revenues to remain on these islands, boosting again the land-based tourism. Between 2000 and 2005, the Galápagos experienced a 71.8% increase in the gross domestic product, from which 74% is attributed to tourism (Mena *et al.* 2013). In addition, currently, the tourism sector employs approximately 40% of the total population of the Galápagos and this represents 65.4% of the local economy.

2.1.3 Socio-economic condition

One of the main causes of migration from Ecuador mainland has been the better employment conditions on the islands. For instance, in 2009, the unemployment rate in Galápagos was 4.9%, while in the mainland of Ecuador was almost 8%. In the same year, underemployment in the Galápagos was 38.7%, while in the mainland was 50.5%, and the full employment was 64.7%, while in the mainland was barely 38%. Because of this, migration from the mainland started to become more and more popular and thus, a very serious problem and threat. According to the latest population census of 2010, the 'economically active population' in Galápagos was around 52% of the total population, whereas in the mainland was 42% (INEC 2010).

In addition, the labour market in the Galápagos Islands has a higher economic chance (70.3%) compared with the mainland Ecuador (67.7%). Also, the salaries tend to be higher, being the double in the public institutions; 10.7% of the population work in the public sector in the islands, which represents the double of the mainland (4.1%). Moreover, more favourable economic conditions are also present in the private sector in the Galápagos Islands. According to a national survey conducted in 2009 by INEC (2010), the minimum monthly income for public and private employees in Galápagos was USD 772/month, while in Ecuador mainland was only USD 251/month. Table 2.2 shows the structure of labour market in Galápagos and Ecuador mainland.

2.1.4 Climate

As of their location on the Equator, the Galápagos Islands should have an equatorial and prevalent warm weather. Nevertheless, the confluence of marine currents and winds, provide these islands with a more temperate weather with two predominant seasons, which contribute highly to the availability or non-availability of water: (I) a hot season 'invierno', which is characterized by hot temperatures from January until May, strong sporadic precipitations, weak winds and high sea temperature (25°C), and (II) a cool season 'garúa', which is characterized by colder temperatures from June until December, frequent drizzle, especially in the higher lands, and presence of strong trade winds and a lowered sea temperature (22.5°C) (Trueman and d'Ozouville 2010).

Analyses of historical meteorological data from the Charles Darwin Foundation showed that the hydrological year in Galápagos runs from June to May (d'Ozouville et al. 2008b, Trueman and d'Ozouville 2010). The so called effective rainfall, the one that mainly contributes for the recharging of the hydrological system begins with the garúa or cool season (June to December) and is minimal in April and May (d'Ozouville and Merlen 2007). The rainfall is highly variable, with recorded monthly total from 0-660 mm in the lowlands and 0-1263 mm in the highlands.

It is greater in Bellavista (located in the highland), than in Puerto Ayora (main touristic centre located in the low lands). This difference is most pronounced during the cool season when rainfall in the lowlands is minimal (Trueman and d'Ozouville 2010).

2.2 The origin of the sources of water

The origin of freshwater in the Galápagos mainly depends on the amount of rain, which contributes to the recharging of aquifers. The variation in quantity of precipitation is defined by three identified factors (CGG 2010): (1) inter-annual factor, (2) geographical factors, such as orientation of watersheds and altitude of the islands, and (3) other factors such as El Nino Southern Oscillation Phenomena which is defined with a return period and impacts, occurring every 2-8 years. This warm phase, El Niño, brings along exceptionally heavy rainfall which creates run-off and recharges the aquifers; the cold phase La Niña brings, on the other hand, drought periods.

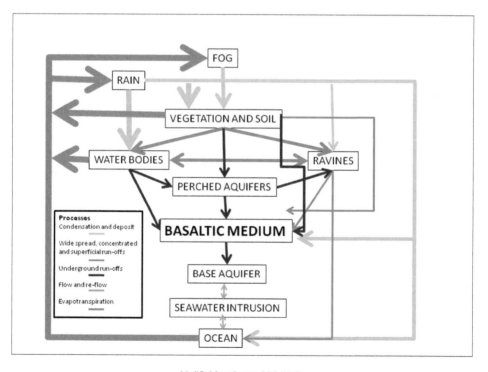

Modified from Source: CGG (2010)

Figure 2.5-Diagram of the Natural Hydrological Cycle in the Galápagos Islands

Water in the Galápagos Islands undergoes several phases and stages in order for it to be available for human consumption as shown in Figure 2.5, the thickness of the arrows approximating the amounts of water. After it has fallen from the atmosphere, rain is captured and intercepted by vegetation and redirected to the soil. Some of the soils are clayey, presenting characteristics of high capacity of water retention (CGG 2010). Moreover, rainwater can directly feed different surface water bodies (lakes, lagoons or wetlands), or transform into ravines. The groundwater feeds ravines of fresh or brackish water, which allows the temporal or permanent storage of water in the soil. For example, Junco Lake in San Cristóbal Island is the only permanent freshwater lake in the archipelago. On the other hand, Santa Cruz has humid zones in the highlands near Cerro Crocker and has some non-permanent lagoons like El Chato lagoon.

Permanent streams exist only in San Cristóbal, which are fed by groundwater. Intermittent streams exist on other islands following very intense rainfalls which transforms into courses such as ravines (activated only in certain times of year), which due to their erosive effect have been able to excavate a fluvial river-bed. The great majority of ravines or streams do not reach the sea (except for a limited number on San Cristóbal), and when they contact the midlands or lowlands with high porosity characteristics, they are quickly infiltrated. The infiltrated water from the upper levels of the system may feed perched aquifers, which may accumulate in a saturated zone. These types of accumulations are produced by the effect of a strong recharge that increases the phreatic stratum, resulting in a space of water retained temporarily by an impermeable level. The perched aquifers exist on San Cristobal and are thought to exist on Floreana and Santa Cruz, as well (CGG 2010).

From the perched aquifers, the water may trespass to the basaltic mean. Several factors affect and determine the circulation and accumulation of water such as geological faults and volcanic dikes. Water may flow vertically through the basaltic medium to the basal aquifer. This aquifer consists of freshwater that has been infiltrated from upper levels of the system, which due to the density difference floats on top of intruded seawater; there is a natural equilibrium between fresh and seawater and usually brackish water is found at the interface. The basal aquifer is influenced by tidal fluctuations, so natural variations in the phreatic stratum are observed as an attenuated tidal signal, combined with the natural recharges of the system and the rate of extraction of the resource. The relation of brackish/freshwater depends on the seawater intrusion, which is a process of natural infiltration of water from the sea into the basaltic mean, and the recharge by precipitations (d'Ozouville *et al.* 2008b).

There are hundreds of watersheds within the islands, each one corresponding to a different type. A watershed refers to an extent of land where surface water from rain or melting snow or ice converges to a single point. Three types of watersheds can be defined on the Galápagos Islands (CGG 2010):

1) Arreic, which means there are no superficial or surface and/or permanent run-offs (Santa Cruz and Isabela).

2) Exorreics are characterized by the flowing on the surface until it reaches the sea, such as the ones found in San Cristóbal Island.

3) Endorreic, have no exit to the sea and the water is contained in a natural storage, such as the Colorada lagoon in San Cristóbal.

2.3 The island of Santa Cruz

Santa Cruz is the central and main island (economic and tourism centre) of the archipelago and has a surface area of 985 km², which extends for 25 km from East to West and for 35 km from North to South. Brackish water is supplied by the Gobierno Autónomo Descentralizado de Santa Cruz (Municipality of Santa Cruz) to the settlements, where the technical capacities continue to be limited. The biggest town is Puerto Ayora, located on the south coast, around Academy Bay, followed by the village of Bellavista, located at 180 m elevation, 7 kms. inland (GADMSC 2012a). Puerto Ayora has approximately 12,000 inhabitants (INEC, 2010), holding 61.3% of the total population of the archipelago and Bellavista has a rural population of approximately 2,500 inhabitants. Over the past 10 years, the rate of development in the highlands has been very high. The growth of some of these precincts is quite recent and poses new water challenges for the municipality, as houses are remote and are not connected to any network, but people do expect services.

2.3.1 Local population and tourism growth

The population density in Puerto Ayora is the highest in the archipelago, with approximately 80 inhabitants per km² (INEC 2010). It is strategically located near the Airport of Baltra Island, built by the American Air Force during the Second World War, which has been bringing visitors from all over the world at an exponential rate since the 1980s. Bellavista is practically a suburb of Puerto Ayora, and it is a popular part of the island characterized by private properties distinguished by tranquility and silence. There are a number of housing developments over the main road going from Puerto Ayora to Bellavista, with less population density.

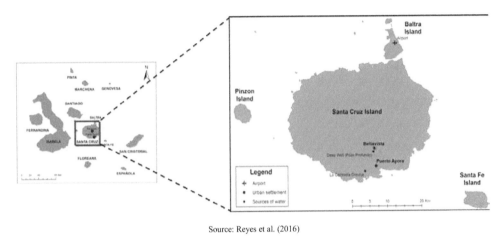

Source: Reyes et al. (2016)

Figure 2.6-Map of the Galápagos Islands, the island of Santa Cruz and the main urban settlements

The access to basic services is still possible through the existing network but is more costly per household than in the concentrated urban areas of Puerto Ayora and Bellavista. Currently, there are new developments taking place further than Bellavista which cause additional problems, as the public network does not reach there yet. Figure 2.6 shows the geographical map of Santa Cruz, indicating the location of the main human settlements.

The main driver for this significant increase in population (from the mainland) over the last years has been to support the tourism industry, as well as an opportunity to generate income. Consequently, there has been a noteworthy increase in the number of travel agencies, restaurants, hotels, bars, etc., additionally stressing the water resources and environment in general (GADMSC 2012b). The growth in the local population and tourism in Santa Cruz has increased the demand for basic services, such as water supply without concurrent ways to cover the costs of these services, overwhelming the municipalities and so resulting in deficient services and untrained staff (Guyot-Tephiane 2012).

Because of this exponential increase, low cost hotels have also increased significantly, as well as the number of backpackers. This has also boosted the creation of small private accommodations or camping sites with no environmental consideration. In addition, more restaurants (also informal) have started, as well as the number of local tourist agencies, offering day tours of occasionally questionable quality. The problem of tourism is further enhanced by the lack of monitoring by the Ministry of Tourism (MINTUR). Consequently, the lack of control and regulation boosts these illegal accommodations. According to the MINTUR, as of to December 2013, there were 106 unregistered accommodations out of a total of 159 (Reyes *et al.* 2017). Moreover, according to the Department of Potable Water and Sewage (DPWS), there are only 40 service connections belonging to the category of hotels (Sarango 2013). In addition, even though the local authorities recognize the problem with lack of regulation of touristic facilities, in August 2014 the Minister of Tourism declared the banning of the moratorium on the construction of new infrastructure for tourist accommodations in the Galápagos Archipelago, in order to exploit these islands as a worldwide tourist destination.

Similar to the tourist accommodations, the number of laundries has grown as well. However, in the land cadastre of Santa Cruz there are no premises categorized as laundries, while according to the DPWS there are only five service connections registered in this category. Evidently, an update of the land cadastre is lacking, as well as more strict control over this type of premises regarding operating licenses. It is commonly known that laundries are a profitable business. Therefore, the total number of these premises is unknown but it is certain that it is much higher than the registered ones.

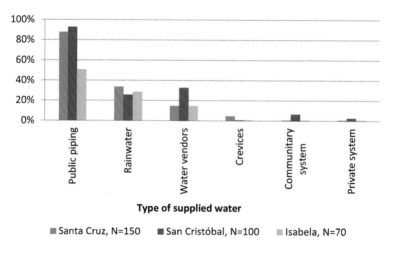

*N refers to the total amount of interviewed people. Source: Guyot-Tephiane (2012).

Figure 2.7 -Percentages of homes supplied by different sources in the three most inhabited islands

As a consequence of uncontrolled urban growth, the local government councils are blocked of providing a permanent and optimal service to the local population. Due to the accelerated growing of water demand, municipalities have shown their incapacity of expansion following current growth rates (Guyot-Tephiane 2012). Consequently, municipal suppliers are forced to ration water distribution, resulting in lack of water supply for at least one third of the population. In addition, in the highlands, rain harvesting no longer covers water demand needs as it used to be, mainly due to evolution of life styles of the population. Figure 2.7 illustrates the different types of water supplied used within the inhabited islands based on a study grounded on interviews to different selected households.

2.3.2 Water supply in Santa Cruz

Unfortunately, the municipal water supply system cannot cope with the current demographic expansion. Among the numerous reasons are financial constraints, lack of personnel and fixed tariff structures as a foundation of the water billing system (Sarango 2013). Also, the volcanic nature of the soil makes expansion of water supply network extremely difficult. Due to these constraints, water is perceived as scarce and the service as poor (Guyot-Tephiane 2012). The water distributed through piped networks has no treatment and is of very low quality. The high concentration of chloride (from 800-1200 mg/L) makes it brackish and not suitable for human consumption. Also, the lack of water metering in the town of Puerto Ayora contributes to unknown water quantities and high water wastage. Several studies have confirmed contamination by *Escherichia Coli* and many water related diseases have been reported (Liu 2011). The aged and unreliable water distribution networks also contribute to the contamination problem. Moreover, the water supply system is intermittent and this has influenced inhabitants to build their own water storage in form of cisterns and elevated tanks. In fact, the roots of these problems seem to be deeper and refer to technical shortcomings as well as issues such as decentralized water supply, lack of consumer's awareness of water conservation and inadequate tariff structure.

2.3.3 Water contamination due to lack of sewage systems

The main operational problem in Santa Cruz is the lack of wastewater recollection and treatment facilities; the domestic wastes are disposed into (precarious) individual septic tanks constructed by the owners, without any technical guideline, affecting water quality severely (Liu and d'Ozouville 2013). As a result, the water quality is the major concern in this island (INEC 2010) due to high concentrations of *Escherichia Coli* found in the water supplied in Puerto Ayora, which has been repeatedly reported since the mid-1980s (Redfern 2003, López and Rueda 2009). The lack of sewage system on the island incentivize disposal systems such as septic tanks, which contribute to groundwater contamination in this highly populated area. Also, septic tanks are not accompanied by a drainage field and are not regularly maintained. The same sewage management systems initially used by the first settlers arriving around the 1930's have persisted through time, regardless of changing realities (Fernández Sánchez 2013).

According to a study made by Liu (2011), very high *Escherichia Coli* levels have been identified also in the storages facilities of the households in Puerto Ayora. While extremely variable from household to household, they were consistently higher than the levels detected at their respective source. Previous studies in other locations around the world have confirmed the patterns of increased contamination in household storages and so along the supply system (Brick *et al.* 2004, Wright *et al.* 2004, Oswald *et al.* 2007)), especially in intermittent supply systems due to the stagnation of water. In conclusion, the contamination occurs both in the distribution system and in the household storage systems (Liu and d'Ozouville 2013). Furthermore, this study made by Liu (2011) also demonstrates that household practices are increasing the level of contamination. In addition, it shows a high correlation between infectious diseases and drinking contaminated water (mainly gastro-intestinal and skin diseases).

Several issues contribute to natural and human contamination: (i) the brackish nature of water, which is not fit for consumption (although early settlers did drink it), so the majority of the population is fully dependent on small private desalination companies (using reverse-osmosis), and a small portion, on rainwater harvesting (d'Ozouville 2009), and (ii) contamination problems affecting the water supply, which refers to lack of sewerage system and treatment, as well as the poor condition of the network (not pressurized and leaks). These reasons have led to the current situation of (contaminated) brackish water supplied by municipal facilities.

The private purification companies, currently six, provide drinking water for homes, offices, restaurants, hotels, boats, and shops (Liu 2011). The majority of water sales occur in Puerto Ayora, with a small percentage going to Bellavista and Santa Rosa. Desalinated-drinking water is sold in three forms: (1) delivered by water trucks (vendors), (2) sold in reusable 20 L containers, and (3) sold in new bottles of 2 L, 1 L and 500 mL. About 30 m^3 of desalinated water is sold daily per purification companiy in Santa Cruz. There is no on-site storage of production since all the desalinated water produced is sold the same day (Liu and d'Ozouville

2013). Even though the bottled water would be expected to be non-contaminated, the results of water analysis by the previous study showed the opposite; water stored in reusable 20 L containers and bottles indicated highly variable levels of contamination with *Escherichia Coli*: 6 CFU/100 mL and 2 CFU/100 mL, respectively.

Until uninterrupted water supply is achieved, re-contamination will likely continue at a high rate and will result in other severe consequences for the water supply system. As mentioned above, even the bottled water occasionally showed non-drinkable quality, which is illustrated in Figure 2.8, showing the variation of contamination by *Escherichia Coli* in different sources and types of desalinated (bottled) water.

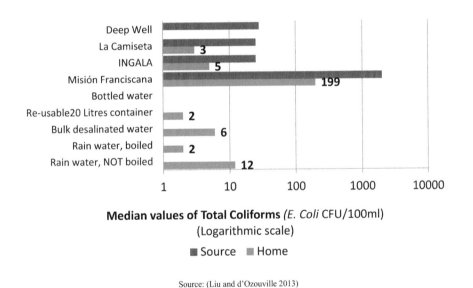

Median values of Total Coliforms *(E. Coli* CFU/100ml)
(Logarithmic scale)

Source: (Liu and d'Ozouville 2013)

Figure 2.8- Source and home comparative contamination levels, showing median values and range of all results

According to the World Health Organization the guideline values for Total Coliforms and Faecal Coliforms in water for human consumption are both 0 CFU/100 mL. In Ecuadorian legislation (Unified Text of Secondary Environmental Legislation - TULAS) the permitted level for Faecal Coliforms is also 0 CFU and for Total Coliforms is 50 CFU/100 mL.

2.4 Research objectives and scope

The main objective of this research is to analyse and quantify current water supply and demand in Santa Cruz Island (Galápagos, Ecuador), generating significant amount of data for a proper improvement in water management in this island. Also, it aims to establish the best and most sustainable solutions regarding mitigation of the water supply system crisis in the next 30 years.

The following are specific objectives of this research focusing on Santa Cruz Island:

(i) To examine quantities of water supplied from each different source (municipal water, bottled-desalinated water, private pumping and rainwater harvesting), and analyse the current state of the water supply system and overall issues contributing to the poor service level.

(ii) To quantify urban water demand for different categories of users (domestic, tourist and commercial) according to the different supply sources.

(iii) To establish the different ranges of domestic water consumption and water demand patterns through water meters' measurements, and analysis of water appliances use in the households.

(iv) To forecast urban water demand and supply on a 30 year planning horizon using the WaterMet^2 model under four population growth scenarios.

(v) To propose different intervention strategies as means of solutions for the water supply and demand crisis, assessing conventional and non-conventional options. Furthermore, to optimize water resources on the island by balancing urban water demand and supply.

(vi) To assess the proposed intervention strategies using an MCDA from different perspectives (environmental, technical, economic and social), considering the standpoints of selected stakeholders and developing specific case-study indicators, in order to find the 'best' alternative.

(vii) To model the hydraulic water supply network, in order to analyse the use of storage facilities in the current intermittent situation and estimate overflow of roof tanks. Additionally, to assess the current water supply network using four different future population growth scenarios.

2.4.2 Key research questions

Based on the above-mentioned specific objectives, the following research questions have been developed, which were addressed and answered throughout the research:

1) What are the different supply sources in the island, the quantities supplied and what is the prevailing situation regarding the municipal supply system?

2) How does the water demand vary among the different categories of users (domestic, touristic and commercial) and the different supply sources? What are the driving forces of demand according to the different sources of supply?

3) What are the ranges of consumption within the domestic sector and what are the water consumption patterns? Is there a correlation between consumption and schedule and zone of distribution established by the municipality of Santa Cruz?

4) What would be the water consumption and coverage of water demand under different population growth scenarios in the next 30 years?

5) What are the most suitable alternatives and intervention strategies that could be used, and to what extent, in order to ensure the coverage of water demand in the future?

6) Which is the 'best' intervention strategy to solve the water supply and demand crisis considering the specific limitations such as availability of water resources and fragility of the ecosystem, the government intentions to increase the revenues from tourism and local stakeholder perspectives?

7) What is the current hydraulic state of the water supply network? Will it suffice for the proposed population growth scenarios under a Demand Driven Analysis (DDA) approach?

8) What is the behaviour of the water supply network under a Pressure Driven Analysis (PDA) approach? Can the storage facilities be modelled in EPANET? What is the estimated quantity of overflow of roof tanks?

This research has scientific, ecological and social significance. Scientifically, it contributes to a better understanding of the current water supply and demand situation in the most inhabited island of the Galápagos Archipelago, generating data and information which previously did not exist. Ecologically, it presents the 'best' and most sustainable option for overcoming water scarcity, considering the fragility and uniqueness of this ecosystem. Finally, it contributes from a social perspective as well, because it analyses consumer behaviour regarding water demand and suggests implicitly towards policy action shifts, aiming for the conservation and better use of the resource, benefitting directly the community, as well as relevant authorities which might be unaware of the real water supply and demand issues.

Our knowledge is a little island in a great ocean of nonknowledge.

-Isaac Bashevis Singer

3

DATA ASSESSMENT FOR WATER DEMAND AND SUPPLY BALANCE ON SANTA CRUZ

This chapter describes the different water supply sources as well as the municipal water supply systems in the main towns of Puerto Ayora and Bellavista. It also elaborates on the problems, the combined influence regarding the lack of data and provides preliminary calculations regarding the supply of water, as well as an estimation of NRW. Also, it explains in detail the perception within the population, as well as the institutional issues that might hinder the optimal management of the water resource.

This chapter is based on:

Reyes, M., Trifunovic, N., Sharma, S., and Kennedy M. (2016). Data assessment for water demand and supply balance on the island of Santa Cruz (Galápagos Islands). *Desalination and Water Treatment* **57**(45): 21335-21349.

3.1 Introduction

The island of Santa Cruz is currently experiencing extreme pressure on its water resources due to exponential increase in tourism and corresponding growth of the local population. Because of that, the municipal water supply system has not been able to provide reliable and constant service of safe drinking water. This chapter presents and analyses the current water supply situation in Santa Cruz by integrating the available information in order to arrive at a sustainable balance between demand and supply. Furthermore, the results of this study will support the implementation of water management measures. Also, this chapter presents a framework to determine the water balance, and portrays preliminary measures for preserving a fragile and unique ecosystem, which may be used in other similar case studies of tourist islands. This is a first step in the direction of determining suitable procedures due to the lack of data, and contributes to a holistic solution by pointing the relevance of water supply and its link with the environment. At a later stage, the aim of establishing a water balance will be to maximize the economic benefits of tourism while preserving the fragile environment.

3.2 Research methodology

A fieldwork was carried out between September 2013 and January 2014 to review the locally available information. Furthermore, several meetings were organised with the main representatives of local institutions to verify the information compiled and investigate whether there was additional work done on water supply.

Table 3.1- Institutions and responsible person selected for the meetings.

Institution	Respondent
Municipality of Santa Cruz	Mayor's advisor
Department of Potable Water	Chief of department
SENAGUA	Chief of department
Ministry of Tourism	Expert responsible for the monitoring program
Charles Darwin Foundation	Chief of the scientific department
DPNG	Chief of applied research department

The purpose was also to identify gaps or overlaps in the information. For this a qualitative method was used, preparing different questions for the different experts, depending on their

Institution	Role	Responsibility	Comments/Remarks
CGG	It is the authority that articulates the regional planning of the islands. Also defines roles and responsibilities of other entities and links them together for the benefit of the islands.	This entity is responsible for developing regulations in order to preserve the islands and control migratory fluxes and illegal immigration.	CGG is also responsible for generating different regulations in order to preserve the islands and control migratory fluxes and illegal immigration.
DPNG	Administers and manages the protected areas, which is 97% of the total territory of Galápagos	This is the supreme authority in terms of conservation within the islands, and is also the responsible for scientific research in order to understand different natural processes and generates lacking data.	Since water is a strategic resource and should be managed by SENAGUA, the national constitution dictates that protected areas and biodiversity are mainly managed by the environmental competent institution, in this case DPNG.
SENAGUA	Highest authority in Ecuador (national level) in charge of managing water patrimony with an integral focus per source (SENAGUA 2012)	Strengthening of the regulations, control of water resources planning and management, private and public concessions and activities that may affect the quality and the quantity of these resources.	Opened operating offices in Santa Cruz only in 2012, with the aim of controlling extraction of water from different crevices, nevertheless, the lack of personnel has made this assignment difficult and long.
Municipality of Santa Cruz	It is the institution in charge of providing some basic services to local population, as well as planning urban expansion and projections of future areas for settlements.	Its department of Potable Water and Sanitation has the competence of providing the water services to the whole island, including the maintenance of the system.	This department is the one in charge of presenting projects for supplying potable water.
WMI	It is a private institution that signed a cooperation agreement with the Autonomous Decentralized Municipal Government of Santa Cruz in 2012	The project is specific for the implementation of optimization and sustainable development of potable water and sanitation systems (WMI-GIZ 2013).	It has financial support of the Deutsch Cooperation 'Deutsche Gesellschaft für Internationale Zusammenarbeit – G.I.Z'

Table 3.2- Main institutions and their responsibilities related to water supply in Santa Cruz

position, knowledge and experience on water issues. This also provided the tool to assess all water related problems from different angles and evaluate different initiatives. Table 3.1 shows the institutions and the respondents contacted during the fieldwork.

3.3 Main stakeholders involved in water resources management of Santa Cruz

Santa Cruz is an area, denominated as a county, with a decentralized autonomous municipality. It is one of the three counties in the archipelago. Water resource management in Santa Cruz is therefore also not centralized.

Table 3.2 summarizes the main water-related institutions within the island of Santa Cruz and their main responsibilities. Several public institutions and entities are involved in the management and/or conservation of water:

1. Secretaría Nacional del Agua (SENAGUA)

2. Dirección del Parque Nacional Galápagos (DPNG)

3. Consejo de Gobierno de Galápagos (CGG)

4. Gobierno Autónomo Descentralizado del Municipio de Santa Cruz

5. Water Management International (WMI) (private institution)

Because of the governance structure, communication amongst the institutions, as well as exchange of information is irregular. The result is occasional overlap between similar studies carried out by various public institutions and NGOs.

3.4 Main suppliers of water at the island of Santa Cruz

There are three types of water supply (suppliers) in the island of Santa Cruz: (a) municipal water supply system, (b) desalinated water from private water purification companies sold in different forms and (c) supply from so called 'private' crevices (boreholes), the latter being out of public control.

3.4.1 Municipal supply

Water provided by the municipality is supplied through two different systems belonging to Puerto Ayora and Bellavista, each one consisting of a different source and a separate distribution network. None of the conveyed water is treated. It is mainly brackish and consequently not suitable for human consumption according to national and international water quality standards. The DPWS of Santa Cruz, reports that 95% of the population of Santa Cruz has access to a centralized water supply system, while the remaining 5% of population have their own supply (from 'private' wells and/or from vendors) (Guyot-Tephiane 2012).

Water from the distribution network is used for most activities except for drinking and cooking. It is regularly used for showering, toilets, and other household activities. The water supply system is unreliable and intermittent, as supply never exceeds three hours per day. Table 3.3 shows the main differences between the supply systems in Puerto Ayora and Bellavista.

Table 3.3- Comparison between water supply systems of Puerto Ayora and Bellavista

Characteristic	Puerto Ayora	Bellavista
No. of connections*	2,591	444
Tariff Structure	Fixed	Metered
Extraction site	Crevice "La Camiseta"	Constructed Deep Well
Type of Water	Brackish (800-1,200 mg of Chloride/L)	Brackish (490 mg of Chloride/L)
Potable treatment (including chlorination)	NO	NO
Extraction rate*	Approximately 3,000 m³/day	Approximately 260 m³/day
Constant Supply	NO	NO
Management	Department of Potable Water	Department of Potable Water

Source: personal communication with the Municipality of Santa Cruz (2013). *Up to December 2013

3.4.2 Desalinated-brackish water

Bottled water is sold to compensate the lack of potable water from the municipal system. The purified water is produced by desalination of brackish water by private companies owing small-scale reverse osmosis plants. There are six desalination companies in the island with unknown production rates (Liu 2011). According to the only possible interview with one of the owners, the average production rate per company per day is around 25 m³ to satisfy demand from local and tourist population. Another study made by Liu (2011), estimates the average daily production of 30.7 m^3 per company. There is no production stocked according to the interviewed person, since the daily production mostly meets the daily demand of the customers fairly accurately. The water is sold in containers of different sizes, and in bulk. This water is mostly used for drinking and cooking, and some for hygiene, depending on the family habits and ability to pay.

The cost of desalinated water is high (2 USD for a 20 Litre container) for it is publicly accepted as the safest water and is the only source of drinking water. According to the recent researches (Guyot-Tephiane 2012), 75% of surveyed homes buy bottled water for human consumption and 67% for cooking, showing the potential for this business. The rest use rainwater or the supplied brackish water. On the other hand, many of the four star hotels have their own desalination plants.

With the approximate population in Puerto Ayora and Bellavista of 14,500 inhabitants, and assuming that all of them buy and consume this type of water, the average daily consumption of desalinated drinking water is 10 lpcpd. This estimate looks high for drinking and cooking purposes, but the water is also sold in bulk to numerous institutions, restaurants and hotels, which are considered to be the major customers of desalinated water. Therefore, the high value per capita can be attributed to the non-domestic consumers.

Table 3.4- Private crevices and their uses throughout the urban settlement of Puerto Ayora.

Name of Crevice	Uses
Misión Fransciscana	Desalination of water for private company.
Tortuga Bay (3 crevices)	Hotels and private properties from Punta Estrada neighbourhood, laundries, etc.
El Barranco (2 crevices)	Private trucks for water sale.
Gallardo	A mechanic place and water desalination company
Martin Schereyer A&B	Own cruises and hotels from owner.
Pampas Coloradas	Private selling with water trucks.

Source: (SENAGUA 2012), (GADMSC 2012a)

3.4.3 'Private' extractions

This source of water refers to brackish water found deep in the crevices emerging from the volcanic origin of the island. Ten registered crevices are located throughout the coastal area covered by the town. The uncontrolled water extraction from these sources is done by pumping;

the exact number of private pumps in each crevice and the quantities extracted is not known, which presents a challenge for the authorities.

Consequently, different owners manage the water resources available on their properties as their own. Some of these private extractions are for personal purposes and others sell and distribute the water by water trucks. Table 3.4 shows the major private water sources with approximate extraction based on interviews and data collected during fieldwork provided by SENAGUA and the Municipality of Santa Cruz. The problem to quantify the total amount of water supplied from the crevices exists since the characteristics of the private pumps are mostly unknown. To illustrate the complexity of the situation, Figure 3.1(a) shows a typical group of pipes from one of the crevices, named 'Misión Franciscana', with no clear indication of the ownership of the different pipes, as well as the quantities extracted. Figure 3.1(b) shows the entrance of this crevice located in the centre of Puerto Ayora.

(a)   (b)

Figure 3.1-Pictures of (a) different pipes and b) entrance of Misión Franciscana Crevice.

3.5 Water supply systems in Puerto Ayora and Bellavista

3.5.1 Water supply in Puerto Ayora

Puerto Ayora water supply system consists of approximately 2,600 service connections (up to December 2013). Its main supply source is from the crevice named 'La Camiseta' located 2.8 km from the urban settlement. Extraction from 'La Camiseta', started in 2011, is done by pumps that supply water to the storage tanks on the site, from where it is distributed to the town by gravity.

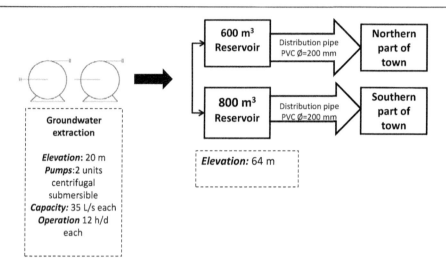

Figure 3.2-Schematic of Puerto Ayora's water supply system

Figure 3.2 shows a schematic of the supply system. The pumping station consists of three 50 HP submersible pumps, (only two working in 2013) (Figure 3.3 (b)). Both pumps extract a total of 70 L/s during 12 h/d and convey water through a 315 mm diameter PVC pipe to two storage tanks (of 600 m³ and 800 m³, respectively), located 2.8 km from the crevice and 64 meters above sea level, as shown in Figure 3.3(a).

(a)   (b)

Figure 3.3-Pictures of (a) Storage tanks and (b) three pumps installed for extraction of brackish water at "La Camiseta".

The tanks are filled during the night and emptied during the day, depending on the schedules of supply, according to the municipality. The average supply is 3 hours per day; however, in some town districts it is 2 hours per day. Two pipelines link the storage tanks with the main network; one is directed to the northern part of Puerto Ayora, and the other to the southern part. Typically, houses in Santa Cruz have their own storage tanks, which vary from house to house (cisterns, elevated tanks, etc.). There is a fixed water tariff structure in Puerto Ayora based on the category of the customer as established by the Municipality of Santa Cruz.

3.5.2 Water supply in Bellavista

Figure 3.4 shows a schematic of the supply system Bellavista. In the case of Bellavista, the water extraction takes place from a constructed deep well called 'Pozo Profundo' (Deep Well). This source is located 200 meters above sea level and has a depth of 160 m, which taps into a basal aquifer.

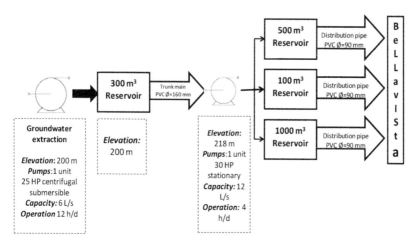

Figure 3.4-Schematic of Bellavista's water supply system

The water is pumped with a 25 HP pump, at 6 L/s, with an average operation of 12 h/d. From there, it is conveyed to a 300 m³ storage tank and from this tank further pumped with a stationary pump of 30 HP with a flow of 12 L/s during 4 h/d, supplied to two reinforced concrete storage tanks (of 500 m³ and 100 m³, respectively) as the last stage before distribution. In 2013, another 1000 m³ tank was built, and all three are now located on the same site, 218 m above sea level. From there, the water is conveyed to different neighbourhoods of Bellavista by gravity. Bellavista has only 444 connections (up to December 2013), corresponding to approximately 2,500 inhabitants. Figure 3.5(a) shows the storage tank and Figure 3.5(b) the extraction pump in Bellavista.

(a)   (b)

Figure 3.5- (a) Storage tank and pumping station and (b) extraction pump at Bellavista

All service connections in Bellavista have a water meter and the water is charged based on the amount consumed, at a price of USD 1.21 per cubic meter. According to the municipality, there have been fewer irregularities observed than in Puerto Ayora, when collecting water bills. Nevertheless, illegal bypass pipes are occasionally installed around the meters to reduce the measurements and the water bills. The service is provided for approximately 2-3 hours per day.

Table 3.5- Differences between two municipal supply systems belonging to Puerto Ayora and Bellavista.

Name	Population size	Pumping Flow (l/s)	Pump Power (HP)	Average pumping (h)	Approximate Leakage*	Extraction (m³/d)	Volume (m³/year)	Water treatment
La Camiseta (Puerto Ayora)	12,000	35 (2 pumps)	50	12	25%	3,024	1,103,760	NO
Deep Well (Bellavista)	2,500	6	25	12	15%	259.2	94,608	NO

*Approximate leakage is according to the Municipality of Santa Cruz

Rainwater collection is more popular in Bellavista than in Puerto Ayora due to higher precipitation levels, and is significant additional water source for people living there (Guyot-Tephiane 2012). Table 3.5 shows the main differences between the two supply systems. The leakage levels are assessed very roughly by the municipality, based on the condition of the networks, and these data would need to be scrutinized. Furthermore, Table 3.6 summarizes the different water supply sources identified in this island.

Table 3.6- Description of different types of water supply in Santa Cruz.

Type of supply	Characteristic of water	Explanation
Municipal (Tap Water)	Brackish-groundwater (400-1200 mg chlorides/litre)	Pumped from crevices by the municipality.
Bottled Water	Brackish- groundwater desalinated by small-private companies with reverse osmosis.	Desalinated from municipal (tap) water or from private crevices.
' Private' pumping	Brackish- groundwater from private crevices	Private pumps and pipes located in various crevices.
Water trucks	Brackish- groundwater	Water pumped usually from private crevices sold by the property owners.
Rainwater	Freshwater	Barely collected in Puerto Ayora, more common habit in Bellavista.

3.6 Water-tariff structures in Santa Cruz Island

The water tariff structure differs considerably between the two supply systems. Based on that, the volume of consumption in Puerto Ayora is not controlled by the consumers, since the same monthly tariff is paid depending on the consumption category. On the contrary, Bellavista has a volumetric water tariff structure, contributing more appropriately to the operations and management of the water supply system. Nevertheless, the established tariffs for both settlements are significantly subsidized by the government, granting a deficit in income to the municipality rather than revenue.

Table 3.7- Water tariffs and number of connections in Santa Cruz.

Setttlement	Category	Number of connections*	Fixed Value (USD)
Bellavista	Metered	444	$1.21/m^3$
Puerto Ayora	Domestic	1156	5.24
	Commercial	944	11.24
	Hotels	14	45
	Industrial/ Laundries	21	45
	Residential	20	28.50
	Official	28	6.12
	Pool	1	28.50
	TOTAL	**2628**	--

*Up to December 2013

Table 3.7 shows different categories for fixed tariffs and the corresponding costs per category, as well as the total number of water connections for each category according to the DPWS

(Municipality of Santa Cruz). Domestic category refers to premises smaller than 100 m² while 'Commercial' are domestic premises larger than 100 m², and also the businesses. The water tariff in Bellavista is also higher due to the quality of distributed water, which is less brackish than in Puerto Ayora (400-800 mg chlorides/litre). The tariff structure in Puerto Ayora presents an immense problem to the municipality because the payment does not vary according to consumption and due to the lack of water metering, the corresponding water demand is unknown. This does not help to create awareness within local population about water conservation despite the regular intermittency.

According to the study by Guyot-Tephiane (2012), customers in Bellavista tend to be more aware in saving water and of its value. Also, because of the volumetric water tariff, rainwater harvesting is a more popular alternative there. Premises in Bellavista have some sort of storage for this type of water, because the precipitation levels are higher than in the lower lands (INAMHI 2014). On the other hand, the same study showed that families in Puerto Ayora do not feel comfortable with rainwater harvesting because they perceive it as an outdated method of water collection. For that reason, they expect local authorities to supply enough water and at affordable prices. Consequently, the lack of awareness in Puerto Ayora is evident due to excessive water wastes, such as spilling of storage tanks.

3.7 Assessment of water consumption for various categories in Santa Cruz

The municipality has been struggling in the last decades to extend the service and improve the supply both in terms of quantity and quality. Different demand categories have been established by the regulations in 2000 and 2004 ('Ordinance which regulates the water service in Santa Cruz County') and later they were reformed in 2005 ('Modified and Complementary Ordinance which regulates the water service in Santa Cruz County'). By law, the municipality has a mandate to fix the tariffs for water consumption and other public services. These tariffs are defined based on analyses of supply schedules for different categories and service areas, and also include maintenance of the system. Nevertheless, the collected revenues do not cover 100% of the operation and maintenance costs; this deficit has to be covered/subsidized by the municipality.

Figure 3.6- Picture showing several elevated tanks within Puerto Ayora.

The exact water demand in Puerto Ayora is still unknown due to the lack of metering. It is therefore difficult to allocate the water demand to different categories of users, except that it is known that the hotels are major consumers within the island. According to an interview at the Ministry of Tourism, in a study made in 2012 by the Spanish Scientific Society called 'Estudio de Subsidio y Huella Ecológica' for Galápagos, the daily consumption of an average tourist was specified at 260 lpcpd. Such figure has been mainly influenced by the construction of swimming pools in the hotels, lately. Therefore, the Ministry of Tourism has developed a project named 'Buenas Prácticas' (Good Practices), where 17 hotels voluntarily subscribed to the plan in order to become more environment-friendly, including the introduction of water saving practices (Giampietro *et al.* 2012). Also, according to an interview at the Ministry of Tourism, several regulations are being imposed to tourist accommodation facilities, where the service can be improved, eliminating low categories hotels and enhancing tourists that can afford higher level of accommodation.

According to the information provided by these hotels, the average expenditure on water per premise is around USD 127 per month. This amount may be considered high when compared to usual water tariffs for hotels of USD 28.50 or USD 45 per month, depending on the size. The average expenditure also considers water bought from water trucks as well as bottled water, since the daily distributed municipal water does not suffice. Furthermore, 10 hotels stated to be interested in saving water while 6 are already purifying their own water via membrane based treatment systems. However, more than 60% of the hotels in Santa Cruz are not legally registered, meaning the authorities are not monitoring these practices in that high percentage of accommodation places.

The water wastages are overwhelming and excessive, especially in Puerto Ayora. The major water losses occur within customer's premises, especially at individual storage systems when they are being filled up and spilling occurs because the faucets are kept open much longer than necessary. This is a common negligence in Puerto Ayora, of which the DPWS is aware of, yet there are no policies regarding this, nor structured system in place to monitor and control it. Figure 3.6 shows the storage coverage at the outskirts of Puerto Ayora.

Centralized potable water and sewage systems have been promised by the Municipality of Santa Cruz for the last 20 years. Nevertheless, the proposed project for potable water has not been finished yet. In 2012, the sewerage pipes were installed in the main (central) neighbourhood of Puerto Ayora, and also around 40% of water pipes were renewed. Yet, the population seems incredulous on potable water supply from the municipality. Even though, as of December 2013, water pipes were restored to conclude finally the potable water component of the project by mid-2015. However, by July 2017 the project is still not concluded, and wastewater recollection and treatment have not been considered yet.

3.8 Management problems in the municipal department of potable water and sanitation

Currently, there are 12 employees in this department. The department is led by a civil engineer with a personal assistant, and 10 staff for technical and administrative tasks. The technical tasks include operation and maintenance of the supply systems and the water meters reading in Bellavista. Furthermore, the employees are responsible for billing, fixing pipe leakages or bursts, general maintenance of equipment and water meters, and cleaning of septic tanks. Obviously, the staff capacity is not sufficient and therefore many activities have to be postponed or skipped.

The limited number of staff as well as the level of expertise is one of the reasons why the municipality is not able to cope with the current population growth, as well as the increase in premises for tourists, restaurants, hotels, etc. The department has stated the need to prioritize their activities, focusing to the repair of bursts in pipes and evident or reported leakages. As a consequence, when the leaks are reported, all other activities are put on halt, leaving the regular work postponed, which suggests that regular maintenance and monitoring are neglected.

The potable water and sewage department also has to cope with the restricted budget. According to the municipality, the current budget is not sufficient for all the activities and expenses of the water supply system due to the old age of the 60% of the network. According to the DPWS, the water tariffs cover only 80% of regular daily operations and management, hence not providing adequate revenue for reliable water service. The deficit of 20% is subsidized by the municipality, generating a significant annual loss which could be covered if the tariffs were higher and differently structured. This economic limitation also portrays a challenge for the water department to expand as expected and provide a better service.

3.9 Institutional issues in Santa Cruz

One of the main problems in Santa Cruz is the lack of governance and policies developed specifically for this type of fragile ecosystem. The different institutions in charge have not yet defined clearly a comprehensive legislation, neither the roles nor responsibilities of different entities involved in water resources. Another issue is the lack of communication and organization among the different entities. Therefore there is an urgent need for policies and regulations regarding water resources management, in order to improve the current situation.

Even though SENAGUA is the lead agency of water resources management in Ecuador, their personnel at Santa Cruz headquarters is also limited. In fact, there is only one person in charge of the office in this island. This person has to deal solely with all the activities and responsibilities, which demonstrates the lack of enforcement and therefore the local population continues to withdraw water from 'private' crevices at their will. Many of land owners assume that they can extract and distribute water without any control, presenting a problem to regulate water concessions of 'private' crevices. Since SENAGUA is relatively new, the population is not aware of the role and responsibility of this institution, therefore showing a certain rejection, and complicating the monitoring and control. Such behaviour makes SENAGUA's work even more difficult. Nevertheless, some managers of private crevices have voluntarily acceded to present the required paperwork so that the institution can monitor periodically the quantities of extraction and the water quality. Furthermore, SENAGUA has not yet developed specific policies or regulations for the Galápagos Islands, regarding the ecosystem's fragility. The regulations that are applied are the same ones as in the mainland, even though the majority of the territory has been declared as National Park, and therefore conservation practices need to be specific.

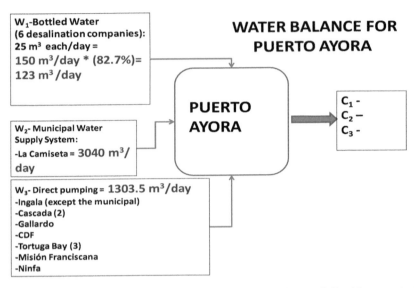

82.7% corresponds to the percentage of Puerto Ayora's population between the two main towns. C_1, C_2 and C_3 correspond to the consumption for each supplier.

(a)

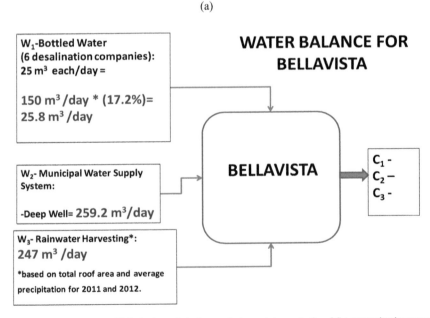

17.2% corresponds to the percentage of Bellavista's population between the two main towns. C_1, C_2 and C_3 correspond to the consumption for each supplier.

(b)

Figure 3.7- Water balance of (a) Puerto Ayora and (b) Bellavista

3.10 Water supply input as component of water balance for Santa Cruz

A water supply assessment in Puerto Ayora and Bellavista was performed based on the information provided by the Municipality of Santa Cruz, as well as from SENAGUA during fieldwork (September 2013 till January 2014). Figure 3.7(a) shows the components on the supply side of the water balance for the supply system in Puerto Ayora and Figure 3.7(b) in Bellavista, based on the different sources. It also reflects the first step towards establishing a water balance by approximating the input volumes into the system.

With the approximations made from available data, an estimated proportion of per capita demand supplied per source type is shown in Table 3.8. The result of 371 lpcpd looks very high but by default it includes leakages; the real consumption would therefore be lower. In the absence of good monitoring, the real leakage levels are yet to be assessed.

Table 3.8- Estimations on total water supply in Santa Cruz Island

Supply source	Quantity supplied (m³/day)	Quantity supplied per capita (lpcpd)
Municipal Water (La Camiseta and Bellavista)	±3283	±226
Bottled (Desalinated Water)	±150	±10
'Private' Extractions	±1951	±134
TOTAL	**±5385**	**±371**

3.11 Water demand calculation in Bellavista

Table 3.9 summarizes the consumption based on the monthly registrations of water meters of the water cadastre for the year 2013 from DPWS. The table shows relatively large number of malfunctioning meters and the consequent calculations of average consumption of water for 2013. As an approximation, the department assumes an average consumption of 1 m³ per month per non-working connection, which seems to be a gross underestimate. However, larger estimated or averaged charged quantities could be easily disputed by the consumers since no evidence is available to the company.

Month	Consumption (m³)	Consumption* (m³)	No. of working devices	No. of non-working devices	% of non-working devices	Average consumption per premise (m³)	Estimated population of non-working devices***	Demand for estimated population of working devices (lpcpd)
January	5,375.6	5,454.6	348	79	18%	15.4	450.3	87.4
February	5,370.2	5,453.2	345	83	19%	15.6	473.1	88.3
March**	330.2	734.2	25	404	94%	13.2	2,302.8	55.8
April**	440.8	847.8	12	407	95%	36.7	2,319.9	81.6
May	4,605.4	4,676.4	358	71	17%	12.9	404.7	73.3
June	6,513.0	6,585.0	360	72	17%	18.1	410.4	103.9
July	6,262.2	6,342.2	353	80	18%	17.7	456	102.1
August	5,559.0	5,641.0	352	82	19%	15.8	467.4	91.2
September	5,653.8	5,742.8	347	89	20%	16.3	507.3	94.6
October	5,653.8	5,743.8	347	90	21%	16.3	513	94.8
November	5,097.8	5,185.8	352	88	20%	14.5	501.6	85.0
December	4,964.8	5,051.8	356	87	20%	13.9	495.9	82.6
TOTAL (2013)	55,826.6	57,458.60	3,555	1,632	—	206.5	9,302.4	1,040.6
Average/ month	4,652.2	4,788.2	296	136	31.5%	17.2	775.2	86.7

*This consumption refers to 1 m³ for non-working devices **Months with high non-working devices due to unknown reasons ***The estimated population was calculated with 7.7 inhabitants per premise.

Table 3.9 - Total water demand, average working and non-working devices and estimation of demand per capita for Bellavista.

As observed in the previous table, the monthly average of percentage of non-working water meters is approximately 32%, which corresponds to approximately 136 meters out of 434 meters that are out of service. This high percentage reflects the significant figures of the months of March and April (which seem to be outliers). The real reasons of this huge deviation are unknown and the municipality did not have valid explanations. As a result, these non-working water meters contribute to a higher value of NRW, which can be considered leakage and/or water theft.

The calculated average consumption per premise per month is ± 17.2 m^3. After some preliminary calculations based on the previous figures, and assuming on average ± 5.7 family members per premise, based on demographic data from 2010 Census (INEC, 2010), the estimated consumption per capita is approximately 87 lpcpd for Bellavista, which can be treated mostly as a domestic use. Furthermore, the per capita demand figure seems to be low, but excludes demand from bottled water, estimated previously at approximately 10 lpcpd. Moreover, with known actual demand and the approximations assuming average consumption for non-registering meters, it was possible to calculate two values of NRW, one for the year 2013 and the other one with a scenario with optimal meter conditions. These two calculated figures help to draw evident conclusions about how the value of NRW could be reduced up to 25% if there would be an improved water-meter management program (Table 3.10). Complementarily, these two percentages give an idea of what could be an estimated leakage percentage, being the difference between the two NRW values.

Table 3.10- Non-Revenue Water with two scenarios (average registering meters and with average consumption for non-registering meters) for Bellavista.

	System Input Volume (m³/year)	Revenue Water (m³/year)	Non-Revenue Water (m³/year)	Percentage of NRW
With average working devices	94,608	±55,827	±38,781	41%
With all devices working	94,608	±80,363	±14,245	15.1%

3.12 Conclusions and Recommendations

In this poor-data case study area, the generation of this data is fundamental for decision-makers. Therefore, this chapter summarises all the current problems affecting the water supply systems in Santa Cruz. Based on the results, it can be observed that the total supply per capita is high, considering the scarcity of water perceived by local population and authorities. However, this figure does not consider leakage yet. Leakage levels, as well as other type of water losses in Puerto Ayora, could be further estimated if specific demand would be assessed and known. Furthermore, water supply figures from the different sources e.g. desalinated-bottled water and 'private' extractions, need to be confirmed, since most likely these figures would be higher.

In the case of Bellavista, it was possible to estimate NRW. The range of values with the two proposed scenarios suggests that the percentage of non-working water meters have a significant influence on this figure, as well as on the total water demand for this settlement. In addition, this also implies that financial revenues from Bellavista supply system decrease to a high extent when water meters do not register any water consumption. Excessive losses and wastage, resulting in high NRW, and lack of proper management, suggests that the population's perception needs to be shifted to a more sustainable one regarding the conservation of water.

Involving the collection of information from different sources and institutions, the lack of communications and cooperation among relevant organizations, influences the current water supply crisis since there is no cross-check of information or follow up of studies. Therefore, this thorough assessment intends to help the authorities seek for specific solutions to the problems portrayed. Also, it suggests that regulations and policies for the conservation of water resources in this fragile ecosystem need to be created and developed, defining exclusive tasks and specific responsibilities among institutions. Since, one of the government's objectives is the exponential growth of tourism, it is necessary to arrive at a full-level scale solution, and not locally-based (as it is currently). Therefore, there is a need for a holistic solution, where the government needs to involve and manage different stakeholders and develop multidisciplinary, interdisciplinary and transdisciplinary approaches.

"Everything must be made as simple as possible. But not simpler."

— *Albert Einstein*

4

QUANTIFICATION OF URBAN WATER DEMAND

This chapter analyses and quantifies water demand in Santa Cruz. The chapter looks at different types of sources, as well as four different categories of consumers. In addition, it compares the domestic demands in the coastal town of Puerto Ayora, the largest in size and extension, and highland settlement Bellavista, which has grown considerably in recent years. Both these settlements have different population size, practices, sources of water, water tariff structures and microclimates. Finally, it attempts a comparison with other islands, which have tourism as a main economic activity. The results are based on the information gathered from the field survey carried out in 2014, combined with information provided by the Municipality of Santa Cruz and other relevant organizations involved in water resources management in the Galápagos.

This chapter is based on:

Reyes, M.F., Trifunovic, N., d´Ozouville, N., Sharma, S., Kennedy, M. (2017). Quantification of urban water demand in the Island of Santa Cruz (Galápagos Archipelago). *Desalination and Water Treatment* **64**: 1-11.

4.1 Introduction

Tourism exerts a significant pressure on water resources in tropical islands. It is the main reason for the increase in water demand, which presents a significant challenge for local authorities to supply water to meet the current growth trends. This chapter analyses thoroughly and quantifies water demand in Santa Cruz, regarding different sources of supply and various categories of which are thought to be major consumers.

The data for this study were collected from a survey carried out with 374 households in Puerto Ayora and Bellavista. Water supply from the different sources was assessed as well, and compared with domestic and tourist water use. The chapter intends to reveal the daily average water demand per capita, and how it differs between two settlements, as well as the impact resulting from difference in the water tariff structures. Also, it analyses the type of consumer in the tourist category, and how much it accounts of the total water demand within the island. Lastly, the chapter elaborates on recommendations on how the implementation of proper WDM measures is essential in order to develop consumers' awareness and sustainable tourism.

4.2 Research methodology

In order to assess the water demand in Santa Cruz, a quantitative survey was carried out during the fieldwork conducted from November 2013 to January 2014 in Puerto Ayora and Bellavista. The minimum sample size was calculated based on the total number of land properties according to the 2012 cadastre from the municipality of Santa Cruz. With a total of 2460 properties in Puerto Ayora and 435 properties in Bellavista, the minimum sample size was calculated at 339, by applying the confidence interval of 95% (DeVault 2014). Next, the actual sample size per consumption category was determined as shown in Table 4.1. Four local assistants were hired to carry out the surveys and 15 residential blocks were randomly selected and assigned to each of them (ten in Puerto Ayora and five in Bellavista), covering a total of 60 blocks. The surveys were carried out during a period of six weeks.

Initially, ten domestic surveys were carried out in order to get feedback of local population on the questionnaire. The selected households found the interview mostly too long becoming indifferent at the moment of answering the questions. This was mostly a result of several similar interviews conducted in the past resulting in little or no improvement of the situation, afterwards. Moreover, several questions regarding the habits and social/economic status were considered sensitive/offensive. Following the trial surveys, the number of questions was reduced and some of the questions were reformulated to make them transparent and culturally/socially acceptable.

Table 4.1-Survey sample size per consumption category in Santa Cruz.

Consumption category	Number of properties	Percentage of total (%)	Optimal number of surveys[a]	Actual number of executed surveys
Puerto Ayora:				
Domestic	1996	69	234	240
Hotels	159	6	19	29
Restaurants	49	2	6	30
Laundries	5	0	1	16[b]
Bellavista:				
Domestic	435	15	51	59
Others (excluded)	251	8	-	-
Total	2895	100	310	374

Note: [a]Calculated according to the procedure at http://www.surveysystem.com/sscalc.htm. [b]Includes not officially registered laundries.

The final version of the domestic survey contained five main parts: (i) general information about the location and description of the household, (ii) family habits addressing the daily routines of the family members, namely the schedules of work, school, preparation of meals, etc. (iii) water demand, referring to the estimates of actual consumption per type of supplied water (bottled, municipal and/or from trucks), (iv) environmental awareness and water saving practices, and (v) sanitation practices, addressing questions related the type of wastewater disposal.

The surveys for other demand categories were less detailed and contained four groups of questions: (i) general information, (ii) average capacity of customers, (iii) water demand quantification regarding different type of sources and (iv) environmental awareness.

4.3 Results of the survey

The results presented in the following section are based on the percentages of responses. Some of the totals do not add up to a 100% because those questions were attempted by choosing more than one of the offered answers.

4.3.1 Water demand analysis in Puerto Ayora

4.3.1.1 Domestic category demand

The survey indicates 92% of the respondents of Puerto Ayora are connected to the municipal network. However, the service is irregular. Figure 4.1 shows the degree of intermittency, where most of the households being supplied a few hours every day.

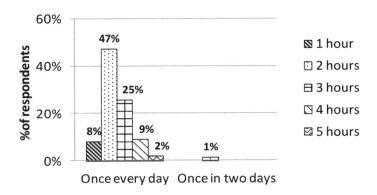

Figure 4.1-Survey results for the inhabitants of Puerto Ayora regarding the percentage of connection and frequency of service.

Figure 4.2- Typical (a) elevated tank and (b) cistern in Puerto Ayora.

Additional water refers to bottled-desalinated water, rainwater harvesting and brackish-water from trucks. 92% of surveyed population claimed to use bottled-desalinated water, while only 8% use rainwater or brackish-water from trucks. Due to evidence of occasional contamination of bottled water, some households reported that they perform additional treatment of this water consisting of filtration, chemical disinfection or boiling. Furthermore, in order to mitigate the intermittency, the households use different types of storage devices, mainly tanks (Figure 4.2 (a)) and cisterns (Figure 4.2 (b)) of various capacities, depending on the family size and habits; and in some cases they even use both.

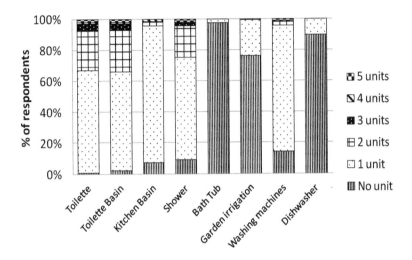

Figure 4.3-Survey results for the inhabitants of Puerto Ayora regarding type and number of water appliances per household.

The survey also considered the type and number of water appliances within a household (toilets, basins, showers, bath tubs, washing machines, etc.). Figure 4.3 shows the number and type of water appliances per household in Puerto Ayora. As observed, the majority of the population does not have a bath tub, garden irrigation system, or a dish washer. Measured with these indicators, a typical family in Santa Cruz is not as wealthy as in some other parts of Ecuador. On the other hand, it may be assumed that residences with more water appliances of the same type likely reflect small-scale tourist accommodation, like rental apartments. The type of water used in the appliances differs. Figure 4.4 shows the percentages of respondents using brackish-ground water and/or bottled-desalinated water for the different household activities.

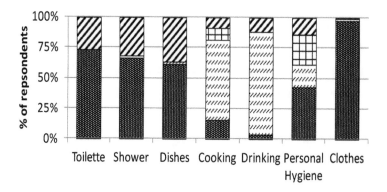

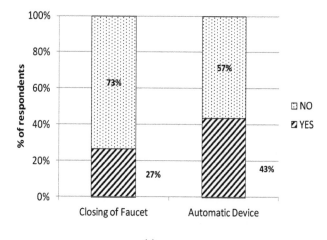

Figure 4.4-Survey results for the inhabitants of Puerto Ayora regarding the use of brackish-ground and bottled-desalinated water for household activities.

Equally important for assessment of water demand, water leakage and spillage of water from individual tanks were also covered in the survey (the latter was observed during the fieldwork). According to the Municipality of Santa Cruz, the greatest losses take place within premises, and result from negligence.

(a)

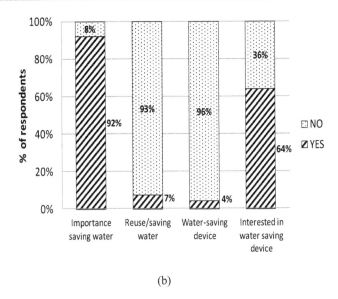

(b)

Figure 4.5-Survey results for the inhabitants of Puerto Ayora regarding (a) negligent practices and (b) environmental awareness in households.

Figure 4.5 (a) shows 73% of the respondents do not close the faucet when the storage tank is filled in their homes and close to 60% do not use float valve or an automatic device to prevent overflow. The often visible result is shown in two examples in Figure 4.6. Although the local population often complain about the lack of service and recognize the importance of saving water, the use of water saving devices such as water efficient toilets or showers is not widespread, although the interest in having one was confirmed by the survey as shown in Figure 4.5(b).

Figure 4.6-Pictures of typical spilling tanks in Puerto Ayora.

Other important information gathered in the survey concerns sanitation practices. 71% reported to have and use a septic tank, while 20% discharged wastewater directly into crevices or the sea. This confirms the contamination of water sources by the proximity of collapsing septic tanks, as mentioned in previous studies (Liu 2011, Liu and d'Ozouville 2013); 63% of the respondents stated that they have never emptied their septic tank, 12% reported cleaning it every two years and only 16% clean it once per year, suggesting that overflowing septic tanks may be contaminating the water sources from the crevices.

4.3.1.2 Tourist category demand (hotels and restaurants)

Several questions were posed about water consumption of various types of tourist accommodations and restaurants. 87% of the hotels and 93% of the restaurants reported that the municipal supply is their main source of water. The volumes of storage tanks and the frequency of filling per week for municipal water employed by hotels is shown in Figure 4.7(a) and in restaurants in Figure 4.7(b).

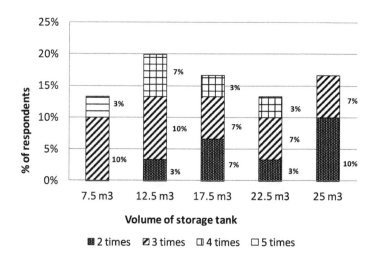

(a)

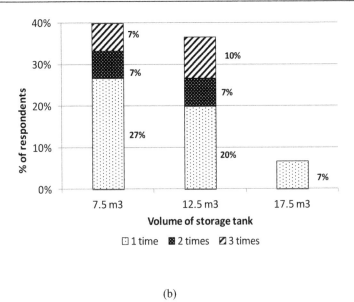

(b)

Figure 4.7-Survey results regarding the frequency of filling of storage tanks per week in (a) hotels and (b) restaurants in Puerto Ayora.

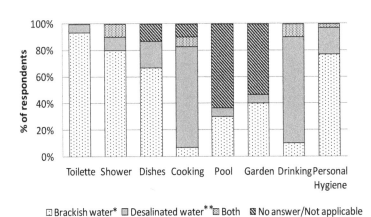

(a)

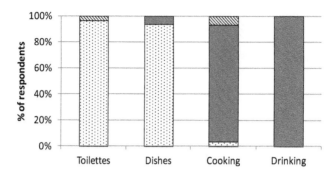

*Indicates desalinated-bottled water or desalination of municipal water and/or trucks. **Indicates water sold by water trucks or municipal water.

(a)   (b)

Figure 4.8-Survey results regarding the use of brackish-ground and desalinated water for (a) hotels and (b) restaurants per type of activity in Puerto Ayora.

In Table 4.2, the average occupancy per type of accommodation (in the case of hotels) and the average number of visitors per day in a restaurant is shown. This information is relevant for further calculation of the total water demand.

Table 4.2-Average capacity and average visitors for hotels and restaurants

Type of hotel accommodation	Average capacity (guests)	Percentage of restaurants	Average number of visitors per day
Hostel	40	20%	<15
2-star hotel	35	20%	20-25
3 star hotel	45	3%	30-35
4-star hotel	35	23%	40-45
		33%	>50

The survey also addressed questions regarding water treatment by hotels and restaurants. It appears that 43% of hotels and 13% of restaurants have their own purification system for the brackish water (from municipal or truck source).

Moreover, there are different types of treatments used in both cases. 61% of hotels use sand filters, 31% use ozonation and 8% use reverse osmosis RO (membrane filtration). The use of water for different activities in hotels and restaurants varies depending on the type of water. Desalinated water (bottled or municipal treated by RO) is used mainly for personal hygiene, drinking and/or cooking. For the rest of activities, such as toilet flushing, dish washing, etc., the brackish-ground water suffices (municipal or from trucks). The results for hotels are shown in Figure 4.8(a) and for restaurants in Figure 4.8(b).

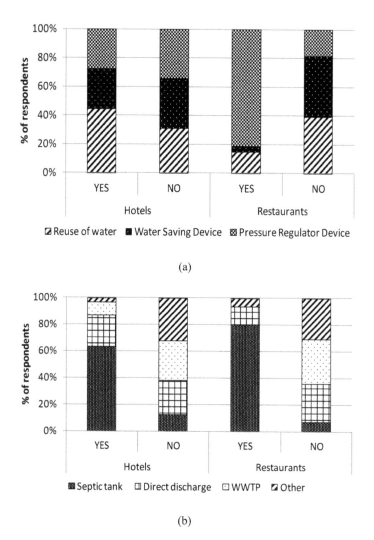

(a)

(b)

Figure 4.9-Survey results regarding (a) environmental awareness in hotels and restaurants and (b) sanitation practices in hotels and restaurants in Puerto Ayora.

Environmental awareness and sanitation is also an important issue to consider among different tourist facilities. Figure 4.9(a) shows water saving practices, comprising the reuse of water,

water saving devices such as water efficient toilets, washing machines, etc., and pressure regulating devices (valves that automatically decrease the pressure and consequently the flow).

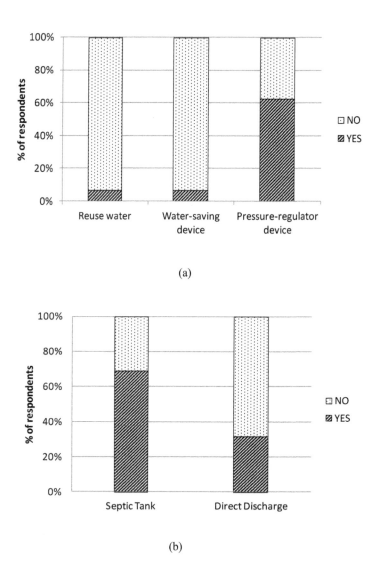

(a)

(b)

Figure 4.10-Survey results in laundries regarding (a) environmental awareness and (b) sanitation practices in laundries.

Concerning sanitation practices shown in Figure 4.9(b), the most common options are septic tanks (63% from hotels and 80% of restaurants), direct discharge/no treatment to a nearby crevice or to the sea (23% of hotels and 13% of restaurants), or own small-scale wastewater treatment plant (10% of restaurants and 13% of hotels). 66% of hotels and 17% of restaurants confirmed that they separate grey and black waters with separate collecting systems.

4.3.1.3 Water demand in laundries

The type of storage (cistern or elevated tank) and the volume used in laundries were also analyzed. 56% of the laundries have storage tanks of size between 16 and 20 m³, 33% between 11 and 15 m³ and 11% between 5 and 10 m³. The volume of the storage tank is related to the number of hours the laundry operates and also to the number of available washing machines. 50% of the surveyed laundries operate 12 or more hours per day, and 57% of these premises have four or five washing machines in operation. Over 63% of the laundries used pressure-regulating devices to lower water consumption, as observed in Figure 4.10(a). Regarding sanitation practices, many laundries dispose their wastewater directly into a crevice or to the sea, shown in Figure 4.10(b).

4.3.2 Water demand analysis in Bellavista

The survey in Bellavista was carried out only for the domestic users, as they are very few hotels, restaurants and laundries in Bellavista. As in Puerto Ayora, the service in Bellavista is also intermittent, with over 80% of the respondents being connected to the municipal network (Figure 4.11). About one third of them receive water only once in three days, portraying a lower level of service than in Puerto Ayora.

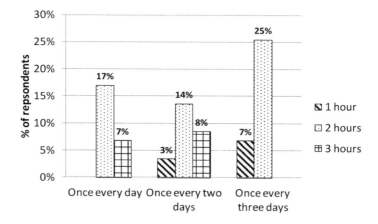

Figure 4.11-Survey results for the inhabitants of Bellavista regarding the percentage of connection and frequency of service of households.

Any shortage of water (since the microclimate is more favourable in Bellavista), is largely compensated with rainwater harvesting (81%). Furthermore, 75% claimed to use bottled-desalinated water and 28% used brackish-ground water from trucks.

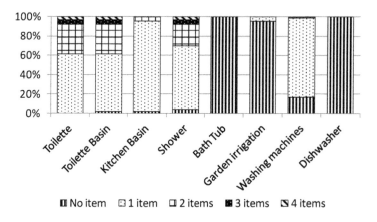

Figure 4.12-Survey results for the inhabitants of Bellavista regarding the type and number of water appliances per household.

Most of the households perform additional treatment, mostly by filtering or boiling water, depending on the type of water used. Due to the intermittency, the use of individual water storage tanks is common practice. The type and the volume vary according to the family size and consumption habits. Cisterns are more frequently used than elevated tanks, and the typical capacity is 3 m^3 per household.

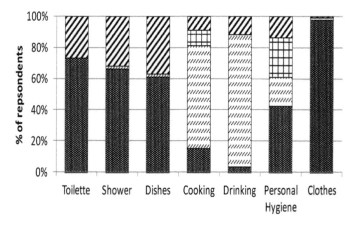

(a)

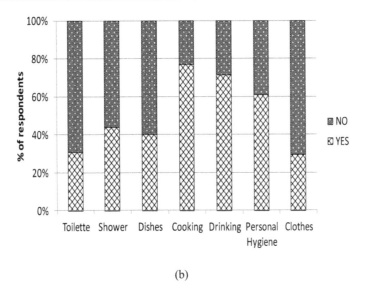

(b)

*Indicates desalinated-bottled water or desalination of municipal water and/or trucks. **Indicates water sold by water trucks or municipal water.

Figure 4.13-Survey results for the inhabitants of Bellavista regarding the use of (a) brackish and bottled water and (b) rainwater for household activities.

Figure 4.12 shows the number of water appliances within households in Bellavista. The situation is similar to Puerto Ayora; where the majority of the population do not have a bath tub, garden irrigation system, or a dish washer. Figure 4.13 shows the type of water used for different activities. As can it can be observed, rainwater is widely used in Bellavista to compensate water, even for drinking (Figure 4.13(a)).

The use of storage tanks was also examined. In Figure 4.14(a) it can be seen that the proportion of households in Bellavista that close faucets when the storage tanks are full, is greater than in Puerto Ayora (53% compared to 27%). Apart from that, reuse of water or the use of water saving devices are hardly practiced in Bellavista as shown in Figure 4.14(b), similar to Puerto Ayora.

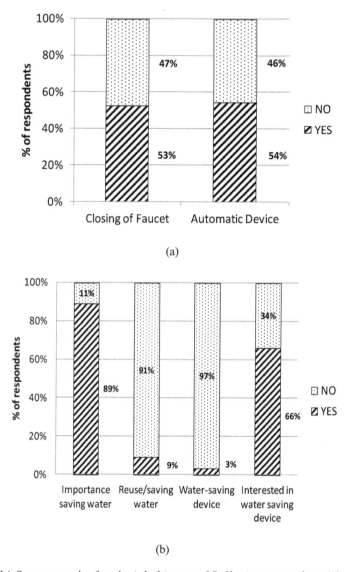

Figure 4.14-*Survey results for the inhabitants of Bellavista regarding (a) negligent practices and (b) environmental awareness at households.*

Sanitation practices in this town are similar to Puerto Ayora. Septic tanks are the most common (88% of the population), and are not cleaned frequently (44% of the households responding to the question have never cleaned their septic tanks and 50% cleaned it once or twice per year).

4.4 Analyses of results in Puerto Ayora and Bellavista

4.4.1 The domestic category

Analysing the results of the survey, several similarities and differences were found after comparing the two settlements. Firstly, the percentage of connections to the municipal service is higher in Puerto Ayora (91%) than in Bellavista (81%). This may be attributed to the faster growth in the number of households in Bellavista and the consequent inability of the municipality to cope with it, since the average annual increase in number of connections from 2005 to 2013 in Bellavista was approximately 9%, while in Puerto Ayora it was only 2%. Furthermore, the frequency of service is clearly worse in Bellavista, as can be seen by comparing Figure 4.1 and Figure 4.11. The response by the households however differs from the information provided by the municipality, who claim that water is supplied every day.

Using information collected on the capacity of household storage units, the reported frequency of filling of storage tanks or cisterns, and the number of users per household, an attempt was made to estimate the total water demand from the municipal supply system, as well as the specific demand per capita. The domestic water demand per capita related to the number of inhabitants per premise is shown in Figure 4.15(a) for Puerto Ayora and in Figure 4.15(b) for Bellavista.

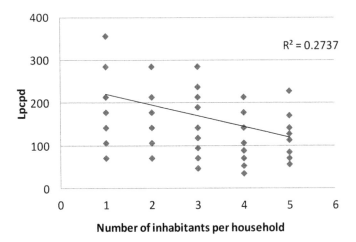

(a)

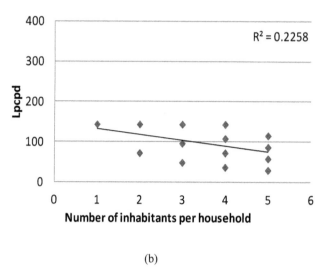

(b)

Figure 4.15-Municipal water demand per capita and number of inhabitants per household in (a) Puerto Ayora and (b) Bellavista.

From Figure 4.15 it can be seen that larger households tend to have lower demand per capita. This can be explained by the fact that the water consumption for general activities like watering gardens, cleaning common areas and cooking is independent of the number of occupants of a household. Furthermore, the figures for Puerto Ayora show a wider range of demand for the same number of inhabitants, suggesting diverse water use, probably due to different living standards and/or habits than in Bellavista. The average specific demand and standard deviation for Puerto Ayora is 163 ± 80 lpcpd. In the case of Bellavista, the average and standard deviation is 96 ± 34 lpcpd. The average water demand per capita (from municipal water) differs significantly between the two settlements, probably due to different water tariff structures.

As stated in Chapter 3, Puerto Ayora has fixed water fees per month for different categories established by the municipality, regardless of the quantity of water consumed. On the other hand, Bellavista has a metered system, with a consumption-based tariff structure, where USD 1.21 is charged per cubic meter. As a consequence, the population in Bellavista tends not to consume as much water as in Puerto Ayora. However, they supplement their demand with rainwater, increasing the total water demand per capita. Since Bellavista has a consumption-based tariff, the customers are more aware of the value of water, unlike in Puerto Ayora, where higher wastage of water is evident with spilling of tanks. The fixed monthly fees in Puerto Ayora seem to be the main reason for such behaviour. According to the municipality, the biggest losses occur at the moment of filling household storage tanks when faucets are not closed when tanks are full, resulting in significant overflow of water. It was observed during the fieldwork that this overflowing tanks were left unattended for more than half an hour in some cases. Lack of metering and low tariffs in Puerto Ayora appear to encourage the population to waste water. However, some households have shown to use much lower quantities of water than the others, for the same number of occupants, which in broader terms reflects different styles of living and/or habits.

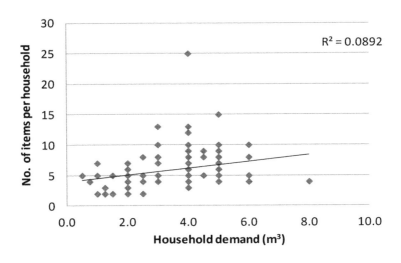

(a)

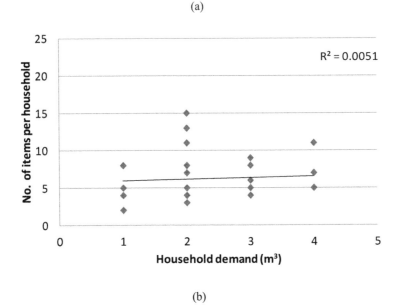

(b)

Figure 4.16-Demand per household and number of water appliances in (a) Puerto Ayora and (b) Bellavista

The high standard deviation on both settlements suggests that locals use water randomly per household, and there is no obvious tendency regarding social stratum or number of occupants, or the neighbourhood. An analysis was carried out to observe if there was a relation between the number of water appliances in the household (toilets, showers, basins, etc.) and the water demand per household. Figure 4.16(a) and Figure 4.16(b) depicts a very low correlation meaning that the demand does not increase necessarily, as the number of appliances increases. The household in Puerto Ayora with a total of 25 water appliances, possibly a tourist accommodation, does not show higher per capita demand, but the reason could also be the limited occupancy of the facility.

A high percentage of responses in Puerto Ayora (32%) identified leaks within their households. Unfortunately, these leaks are rarely fixed, probably due to the fact that the water lost is not charged to the consumer. On the other hand, the leaks within premises in Bellavista are much lower (reported in 15% of responses), meaning that the tariff structure influences the decision to fix them. The majority of leaks were reported to be in old and inefficient toilets, suggesting high losses since toilets account for major water use in a household of nearly 30% of total water consumption (EPA 2012). Finally, the less frequent spilling of individual tanks in Bellavista confirms overall higher awareness of the customers, which can be also explained by the difference in water tariff.

4.4.2 Total demand quantification for domestic sector

The quantification of domestic water demand was done for all types of sources used. This was possible based on the questions regarding the frequency of service/purchase and volumes of the different sources for every household. The results are shown in Table 4.3 with total demand per capita calculated for both settlements.

Table 4.3-Water demand per different sources of water in Santa Cruz

Settlement	Municipal supply (m³/year)	Bottled water (m³/year)	Water trucks[a] (m³/year)	Rain water[b] (m³/year)	Total demand (m³/year)	Approximate population (no. Inhabitants)	Specific demand (lpcpd)
Puerto Ayora	712,188	7,243	57,518	N/A*	776,949	12,000	177
Bellavista	82,481	2,683	48,307	97,444	230,914	2,500	253
TOTAL	794,669	9,925	105,825	97,444	1,007,863	15,000	190

Note: [a] Water from trucks refers to partial pumping from ' private' crevices. [b] Rainwater was not considered in Puerto Ayora for it is practiced by less than 10% of surveyed households.

Table 4.3 indicates relatively high specific demand in view of the widespread intermittency and scarce water sources. The public perception in both settlements is clearly that additional water next to that supplied by the municipality is necessary. Nevertheless, rainwater is barely collected in Puerto Ayora. One reason is lower precipitation than in Bellavista, but also that this practice is considered archaic (Guyot-Tephiane 2012). Oppositely, people in Bellavista collect rainwater regularly and use it for all household activities. Furthermore, in both settlements the bottled-desalinated water is used mainly for drinking and for personal hygiene, while brackish ground water is used for other domestic activities such as cooking, dish washing, laundries, toilet flushing and showers.

Table 4.4-Domestic water demand per capita in various tropical Islands.

Island	Village	Demand per capita (lpcpd)
Crete (Greece)	Chania	443
	Rethimno	424
	Iraklio	474
	Lasithi	338
Aegean Islands (Greece)	Lesvos	246
	Chios	203
	Samos	270
	Dodecanese	207
	Cyclades	164
	Kalymnos	285
Santa Cruz de Tenerife (Spain)	Tenerife	250
Mallorca (Spain)	-	300
Korčula (Croatia)	-	184
Barbados	-	209
Jamaica	-	160
Dominican Republic	-	421
Trinidad & Tobago	-	324

Adapted from (Kechagias and Katsifarakis 2004, Ekwue 2010, Hof and Schmitt 2011, Konstantopoulou *et al.* 2011, Bonacci *et al.* 2012, Guilabert Antón 2012)

Moreover, the supply by water trucks has high contribution to the high total demand in Bellavista, which could be explained by lower number of municipal service connections. In summary, the average per capita consumption is considerably higher in Bellavista than in Puerto Ayora. It is however to be noted that all the results are based on the personal assessments of the respondents; this certainly needs to be verified by more accurate measurements. Currently, there is an ongoing study where water metering is implemented in a pilot section in Puerto Ayora, which will help to validate the figures obtained from the survey.

The calculated domestic demand in both settlements is comparable with the domestic demand reported in the literature on other tropical islands dealing with tourism. Nevertheless, Table 4.4 shows the average for Santa Cruz in Galápagos to be lower than in Greece (Crete, Aegean Islands), Spain (Mallorca), Dominican Republic and Trinidad & Tobago.

Table 4.5-Non-Revenue water estimation for Puerto Ayora and Bellavista.

Settlement	Demand (lpcpd)	Total Demand (m³/year)	Total Supply (m³/year)	NRW (m³/year)	Percentage (%)
Puerto Ayora	163	712,188	1,103,760	391,820	35
Bellavista	96	87,600	94,608	7,008	7

4.4.3.1 Estimations of NRW

Based on the municipal supply and the demand calculated from the surveys for both settlements, NRW estimations were made as shown in Table 4.5. Additionally, for the case of Bellavista, other scenarios have emerged along the analyses of obtained data by using the water cadastre. The NRW estimates vary for different scenarios as shown in Table 4.6. The figure calculated from the official cadastre information provides a high NRW value, which indicates large economic losses for the municipality due to 32% of malfunctioning water meters.

Table 4.6-Different water demand and NRW scenarios for Bellavista

Scenario per source of information	Demand (lpcpd)	Total Demand (m³/year)	Total Supply (m³/year)	NRW (m³/year)	Percentage (%)
Surveys	96	87,600	94,608	7,008	7
Municipal Cadaster*	56	50,862	94,608	43,746	46
Municipal Cadaster (average calculation)**	87	79,130	94,608	15,478	20

*Based on the average consumption from water meters' readings **Based on an average calculation assuming malfunctioning water meters will register an average consumption

4.4.3 Analysis of tourist and laundry category

The total demand was also assessed for tourist facilities in Puerto Ayora: private apartments, hotels and restaurants as shown in Table 4.7. The figures have been derived based on the survey questions regarding the volume of storage facilities and the frequency of refilling storage tanks, as well as amount of bottled-desalinated water and water supplied by trucks.

Table 4.7-Water demand quantification for hotels and restaurants in Puerto Ayora

Type of accommodation	Average capacity (customers)	Municipal water (m³/day)	Water trucks (m³/day)	Bottled water (m³/day)	Specific demand (lpcpd)
Hostel	40	8.1	0	0	205
2-star hotel	35	4.0	12.3	0.1	470
3 star hotel	45	6.0	29.7	0.3	667
4-star hotel	35	9.6	9.0	0.1	535
AVERAGE	**38**	**7.0**	**11.3**	**0.1**	**469**
Restaurants	15	0.2	0.9	0.1	126
	25	0.5	1.7	0.1	158
	45	0.4	0.9	0.2	46
	50	0.4	1.8	0.3	79
AVERAGE	**34**	**0.4**	**1.3**	**0.2**	**102**

Figure 4.17 (a) shows the daily demand of surveyed hotels; the horizontal axis represents each hotel, given by a serial number. For example, from 1 to 6 are hostels with capacity of 25 to 45 people, from 7 to 9 are two stars hotels with capacity of 20 to 25 people, from 10 to 26 are three stars hotels with capacity of 10 to 50 people, and from 26 to 30 are four stars hotels with capacity of 25 to 50 people. The water demand varies according to the type of accommodation (hotel rating) and the average capacity. Additionally, the hotels reported either they have their own purification systems, or they perform additional treatment for the municipal water as shown in Figure 4.17(b).

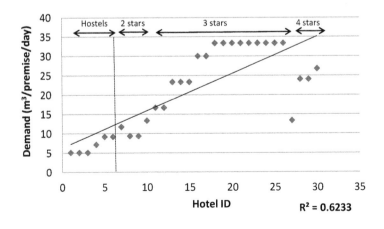

(a)

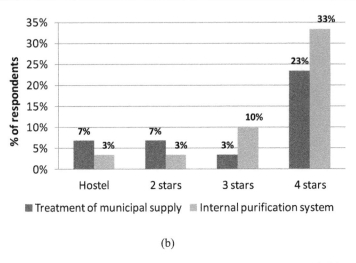

(b)

Figure 4.17- (a) Demand per surveyed hotel regarding rating and (b) internal purification per surveyed hotels.

The majority of the hotels and restaurants are connected to the municipal supply, but some hotels (mainly three stars) and virtually all restaurants are mainly supplied by water trucks. The four stars hotels mostly have their own purification systems (by desalination) and are less dependent on the municipal supply, using it less than lower class tourist accommodations do. Moreover, three star accommodations use more water because of higher occupancy. In Galápagos the tendency is towards middle class tourists; those who cannot afford luxurious accommodations (at average 350 USD/night) which seem to be more careful with water use.

The average water demand per bed, estimated by the survey respondents was 168 lpcpd, which is far from reality, since the calculated average is 469 lpcpd, i.e. almost three times higher. Furthermore, in ten out of 30 restaurants the water demand estimate was approximately 0.5 m³ per day.

Table 4.8-Total water demand quantification considering all categories

Category	Municipal supply (m³/day)	Bottled water (m³/day)	Water trucks[a] (m³/day)	Total demand (m³/day)
Domestic	1,951.2	19.8	157.6	2,128.6
Hotels	1,107.2	20.6	1,788.8	2,916.6
Restaurants	69.3	7.6	51.1	128.0
Laundries	28.5	0	20.1	48.6
TOTAL	3,156.2	48.0	2,017.6	5,221.8

Note: [a] Water trucks refers to pumping from 'private' crevices.

Finally, the total water demand for Puerto Ayora was calculated based on the average consumption derived from the survey, multiplied by the total number of premises per category according to the land cadastre of the municipality. Table 4.8 shows the highest demand from the hotels for municipal water and from trucks. Approximately 24% of total demand of municipal water is from unregistered accommodations, which account for 66% of the total in the category of hotels. As in many other tropical islands, the biggest consumers are tourist accommodations, proportionally to the ranking of hotel. The restaurants and laundries do not contribute significantly to the total demand in Santa Cruz, although the total number of registered restaurants and laundry premises in the land cadastre may be higher than reported.

4.5 Average costs of water supply in Santa Cruz

Based on the information on average demand per category in Puerto Ayora, total revenues per category for the municipality are shown in Table 4.9. This table shows revenues from the water cadastre 2013 and an estimated actual price per cubic meter, based on average number of connections. As observed, revenues from fixed tariff structures in Puerto Ayora are significantly low, considering the actual volume of consumption of water. The estimated price paid per cubic meter for all categories is low as well, considering the minimum salary for the islands, which is approximately 700 USD/month. Therefore, the current payment of a water bill for a family receiving just one minimum salary would represent only 0.8% of the monthly income.

Table 4.9- Average revenues per month and per category in Puerto Ayora for the year 2013.

Category	Average number of connections	Fixed Value (USD)	Average revenue (USD/month)	Average consumption per premise (m³/month)	Average cost of water (USD/m³)
Domestic (less than 100 m²)	1146	5.24	5 716	16.2	0.31
Domestic (more than 100 m²)	886	11.24	10 275	18	0.61
Commercial (restaurants)	49	45	162	42.4	0.26
Small hotels	21	28.50	917	182.9	0.24
Big hotels	20	6.12	558	235	0.12

On the other hand, Bellavista is mainly considered domestic, since there are very few premises belonging to other categories. Since around 32% of water meters do not work properly (as reported in 2013), the registered total consumption is significantly lower. Table 4.10 shows the average consumption per premise based on working meters, then compared to the collected revenues in order to calculate actual price of water per cubic meter. Table 4.10 also shows that average actual payment per cubic meter is approximately USD 1.05, explained by the high percentage of non-working meters. Clearly, this issue contributes to extra financial burden to the municipality, since the expected revenues are even lower (connections with non-working meters are only charged USD 2.21 per month). These calculated values are considering theoretical revenues only, not taking into account yet the overdue bills.

In the survey, an inquiry was made about the actual price people pay for water. As shown in Figure 4.18, 28% of the population in Puerto Ayora and 17% in Bellavista do not pay any water tariff at all. This matches with a further interview made to the DPWS, in which they confirmed that they do not suspend the water service in Puerto Ayora to premises that do not pay. They also explained that the department would need to break the streets and dig in order to suspend the service because there is no valve for each connection to shut it off. Due to lack of personnel and financial resources, this is hardly done. In theory, the penalty is suspension of the service after two months of no payment and an extra fee of USD 6 for the reconnection of the service, action which takes place only in Bellavista due to the presence of water meters. Nevertheless, this policy is not applied in Puerto Ayora and in reality there is no penalty for lack of payment of the monthly tariffs.

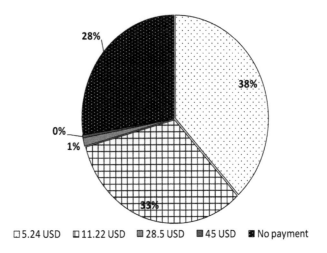

(a)

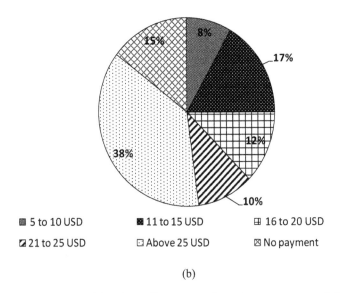

| ▨ 5 to 10 USD | ▨ 11 to 15 USD | ⊞ 16 to 20 USD |
| ◪ 21 to 25 USD | ▫ Above 25 USD | ⊠ No payment |

(b)

Figure 4.18- (a) Payment of fixed tariffs per month in Puerto Ayora and (b) average payment of monthly water bills in Bellavista

Furthermore, the overdue bills are an important obstacle for the municipality, especially in Puerto Ayora, where customers can not be disconnected to the lack valves. In 2013, the number of customers who did not pay on time increased by approximately 15%, as shown in Figure 4.19. In the case of Bellavista, the percentage of overdue bills is significantly lower and showed the trend of decrease following the end of the year; tendency that can be attributed to metering.

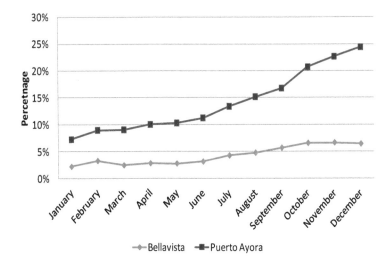

Figure 4.19-Increase of overdue bills for the year 2013 in Puerto Ayora and Bellavista.

Month	Registered Consumption (m³)	No. of water meters registering consumption	No. of meters not registering consumption	Average consumption/ premise (m³)	Real calculated consumption (m³)*	Total billed (USD)	Price of water (USD/m³)
January	5,376	348	79	15	6,596	6,931	1.05
February	5,370	345	83	16	6,662	6,926	1.04
March**	330	25	404	13	5,666	829	0.15
April**	441	12	407	37	15,391	952	0.06
May	4,605	358	71	13	5,519	6,002	1.09
June	6,513	360	72	18	7,816	8,313	1.06
July	6,262	363	80	18	7,681	8,010	1.04
August	5,559	352	82	16	6,854	7,160	1.04
September	5,654	347	89	16	7,104	7,277	1.02
October	5,654	347	90	16	7,120	7,278	1.02
November	5,098	352	88	14	6,372	6,608	1.04
December	4,965	356	87	14	6,178	6,450	1.04
AVERAGE	5,506	352	82	16	6790	7096	1.05

*Consumption calculated assuming all non-working devices will register the average consumption for that month **These months were excluded from all average calculations since they do not represent a typical month.

Table 4.10-Actual price of water based on all water meters working in Bellavista.

The figure implicitly explains the lack of proper management since the population appears to be increasingly encouraged not to pay in the absence of punitive measures. The rate of expansion of the settlement due to tourism growth, and the increase in number of water connections without any control measures will only contribute to increase this trend.

Table 4.11- Financial deficit for Puerto Ayora and Bellavista

Settlement	Cost of supplied water* (USD/year)	Total billed (USD/year)	Total collected (USD/year)	Deficit with total billed (USD/year)	Deficit with total collected (USD/year)
Puerto Ayora	993,384	211,538	190,926	781,846	802,458
Bellavista	114,476	74,744	71,620	39,732	42,856
TOTAL	1,107,860	286,282	257,653	821,578	850,206

*Considering only operations and management costs

Some further calculations and analysis based on costs of abstraction of water and bill emissions and collection, estimated the total revenues for the municipality. The figures for total revenue collected were calculated subtracting overdue bills and lack of payments (The figure implicitly explains the lack of proper management since the population appears to be increasingly encouraged not to pay in the absence of punitive measures. The rate of expansion of the settlement due to tourism growth, and the increase in number of water connections without any control measures will only contribute to increase this trend.

Table 4.11). The cost of supplying water includes only operation and management costs for both settlements, and excludes a significant financial investment done by the municipality (Personal Communication, 2014).

4.6 Analysis on willingness to pay, payment of bottled water and increase of fixed tariffs

The water service provided by the municipality is low, and as a consequence the tariffs have been established with a low price. In a personal communication with the municipality, they affirmed not to increase prices of water due to a fear of rejection. Nevertheless, respondents from the survey said they are willing to pay more, conditioned to receiving a better service and better quality of water. The results of these affirmations are portrayed in Figure 4.20.

Figure 4.20(a) shows that more than 80% in Puerto Ayora and more than 60% in Bellavista are willing to increase their monthly payments in exchange for a more reliable service, as well as a better quality of water. Figure 4.20(b) illustrates that more than 60% of the surveyed population in Puerto Ayora affirm to be willing to pay between 10 to 20 USD per month and around a 30% between 20 USD and to 30 USD. In Bellavista, around 40% of the population is willing to pay between 10 to 20 USD per month, while the same percentage is willing to pay between 20 to and 30 USD per month for an improved service. This suggests, in fact, that local population is aware that a better supply system will require an increase of current tariffs.

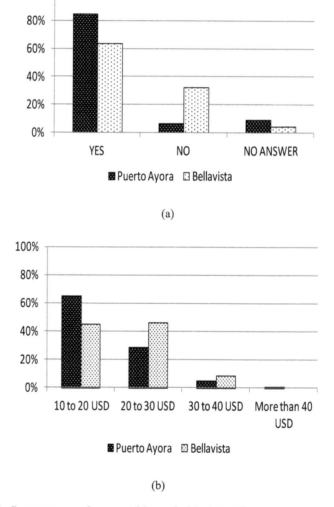

(a)

(b)

Figure 4.20- Percentages of surveyed households (a) willing to pay more for a better municipal service per month and (b) the amount willing to pay per month.

It is also important to analyse the actual total payment of water. These costs reflect the consumption of bottled-desalinated water, which is considered expensive. For example, in

Puerto Ayora, 46% pay between 5 and 10 USD per month for drinking (bottled-desalinated water). In Bellavista, the majority of households surveyed pay more than 20 USD per month for bottled water. The results of the monthly payment for bottled water are shown in Figure 4.21.

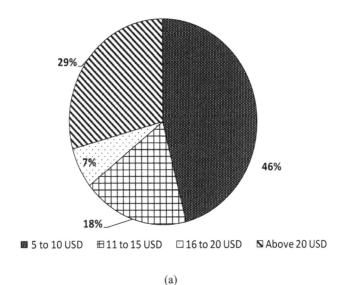

(a)

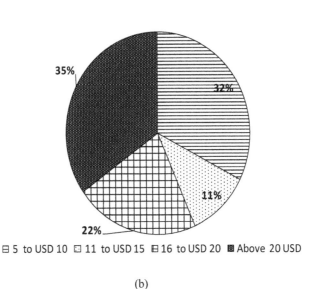

(b)

Figure 4.21- Average payment per month for bottled water in (a) Puerto Ayora and (b) Bellavista

Figure 4.21 shows the distribution of monthly expenses per family for bottled water among surveyed population. This indicates that an average family pays significantly more for all water sources. Based on their willingness to pay and the actual payment, it is possible to create several scenarios to analyse the impact of increase in the revenues for the municipality as shown in Table 4.12.

Table 4.12- Various scenarios on increase of current water tariffs

Settlement	Total Billed for 2013	Scenario 1 (20%)	Scenario 2 (40%)	Scenario 3 (60%)
Puerto Ayora	211.538	380.768	528.844	634.613
Bellavista	74.744	126.797	174.082	207.858
Total	**286.282**	**507.565**	**702.927**	**842.470**
Deficit 1*	*2.892.547*	*2.671.264*	*2.475.902*	*2.336.358*
Deficit 2**	*821.578*	*600.295*	*404.933*	*265.389*

Source: Water Cadastre of 2013 of the Municipality of Santa Cruz. *Includes investment costs **Only operations and management costs

As observed in Table 4.12, even when the tariffs would increase by 60%, the deficit for the municipality does not decrease significantly, as expected. In the case of the second deficit, the cost cannot be covered even when the water bills are increased by 100%. This fact suggests that water tariff structures need to be completely reformulated. However, in order to increase the tariffs drastically, the water supply service would need to improve proportionally. Since the municipality has limited means to improve the service, due to limited revenues, the situation is in vicious circle.

4.6.1 Scenario with Linear (Volumetric) Tariff

Based on the results from the previous section, increasing current tariffs up to 60% would not suffice for the municipality to increase significantly their revenues. Therefore, other tariff structures have been proposed. In Table 4.13, a linear tariff is suggested, where payments are directly proportional to consumption. The cost per cubic meter was assumed as the same current price as in Bellavista (1.21 USD/m^3). Also, an investment of water meter installation of 151 USD per water meter was considered (Personal Communication, 2014). Furthermore, based on results on specific demand from the different categories from the survey, published in Reyes *et al.* (2015), the results on revenues for the municipality are as shown also in this Table. These averages were calculated based on the estimation of surveyed people on the volume of their storage tanks and the times of filling per week.

If tariff is changed to the same scheme as in Bellavista, the revenue for the municipality will increase significantly. However, an investment for water meter installation is needed and this will decrease somewhat the revenues once more. Nevertheless, in the following year the revenues would increase in a more significant way.

Table 4.13-New calculated revenue for the municipality with Linear Tariff Structure

Category	Average consumption per premise (m³/month)	Average no. of connections	Corrected number of premises*	Revenue with Linear Tariff (USD/year)
Domestic (less than 100 m²)	16.2	1145.9	1,146	269,666
Domestic (more than 100 m²)	18.0	886.0	443	115,629
Commercial (Restaurants)	42.4	49.0	492	303,103
Small Hotels	182.9	20.5	80	212,407
Big hotels	235.0	19.8	80	272,927
Bellavista	15.0	444.0	444	96,703
			TOTAL	1,270,436
			WATER METER INVESTMENT	338,391
			TOTAL REVENUE	932,045

*Refers to estimation of what could be the actual number of connections per category, since the average number according to the municipality is not accurate.

4.6.2 Scenario with implementation of Increasing Block Tariff (IBT)

With an IBT structure, major consumers would pay more, especially at the higher blocks of consumption. Table 4.14 shows a potential IBT structure, where the base tariff refers to a fixed cost to any consumption within the range of the first block. The following blocks reflect the cost per cubic meter after the base tariff has been exceeded. Therefore, when the consumption increases, so does the cost per cubic meter.

The base tariffs for the fists block for the major consumers start with higher costs because these categories account for more than half of total demand (Reyes *et al.*, 2015). The average payment per premise per category was calculated based on the average demand per premise shown in Table 4.13. The costs selected are similar to the ones already applied in Bellavista, so it is thought that there would not be an excessive rejection. The new revenue for the municipality with an IBT structure was calculated, considering the same water meter installation investment as the previous scenario.

Table 4.14- Suggested Increasing Block Tariffs in Puerto Ayora

	Ranges (m3)	Cost (USD)	Average payment per category
DOMESTIC			
Base tariff	*1-8*	5	(less than 100 m²) = 14 USD/month
			*Based on average demand of 16.2
Block 1	>8-15	1.1/m³	m³/premise
Block 2	>15-20	1.3/m³	(more than 100 m²) = 16.6 USD/month
Block 3	>20-25	1.5/m³	*Based on average demand of 18
Block 4	>25-30	1.7/m³	m³/premise
COMMERCIAL			
Base tariff	*1-10*	8	
Block 1	>10-20	1.1/m³	
Block 2	>20-30	1.3/m³	50.4 USD/month
Block 3	>30-40	1.5/m³	*Based on average demand of 42.4
Block 4	>40-50	1.7/m³	m³/premise
RESIDENTIAL (Small hotels)			
Base tariff	*1-20*	15	
Block 1	>20-60	1.3/m³	
Block 2	>60-120	1.5/m³	270 USD/month
Block 3	>120-150	1.7/m³	*Based on average demand of 182.9
Block 4	>150-200	1.9/m³	m³/premise
INDUSTRIAL (Big hotels)			
Base tariff	*1-30*	25	
Block 1	>30-80	1.4/m³	
Block 2	>80-150	1.6/m³	356 USD/month
Block 3	>150-220	1.8/m³	*Based on average demand of 253
Block 4	>220-300	2/m³	m³/premise

Table 4.15- New calculated revenue for the municipality with IBT.

Category	Average payment per premise (USD/month)	Average number of connections	Corrected number of connections*	Revenue with IBT (USD/year)	Revenue with IBT and corrected number of connections (USD/year)
Domestic (less than 100 m²)	14	1,146	1,146	192,514	192,514
Domestic (more than 100 m²)	16.6	886	443	176,491	176,491
Commercial (*Restaurants*)	50.4	49	492	29,635	297,562
Small hotels	270	21	80	66,592	258,248
Big hotels	356	20	80	84,728	339,624
Bellavista	12.7	444	444	67,666	67,666
				TOTAL	**1,332,104**
				WATER METER INVESTMENT	**338,391**
				TOTAL REVENUE	**993,713**

*Refers to estimation of what could be the actual number of connections per category, since the average number according to the municipality is not accurate.

The total revenue for the municipality was calculated based on the average payment per premise for each category, which was based on average consumption per household and the number of connections for each category. Table 4.14 presents the suggested increasing block tariff for Puerto Ayora. Since hotels are the major consumers and account for 55% of the total water demand, this tariff structure seems to fit, making major consumers to subsidize the lower consumers. With such an increase, the municipality will almost cover operations and management costs, including the investment for water meter installation (Table 4.15). It is important to mention that the even at this tariff water supplied would still be of non-drinking quality, but will give the municipality to increase their revenues and have the capital to improve the quality with different treatments.

4.7 Conclusions and Recommendations

Water demand quantification for the island of Santa Cruz has been challenging; the available data are scarce and inaccurate. Nevertheless, the conducted survey provided a scenario reasonably close to reality and has enabled preliminary calculations of water demand for different consumption categories. The highest water demand was observed for the hotels, with an average consumption of 469 lpcpd, which accounts for 49% total water demand on this island. This figure complies with the literature data on water consumption for other tourist accommodations on tropical islands worldwide, who are threatened also by high tourism growth rates.

Regarding the domestic water use, there is an evident difference in per capita averages of consumption of municipal water (163 lpcpd in Puerto Ayora and 96 lpcpd in Bellavista) suggesting that different tariff structures may influence the consumption. In Bellavista, where water is charged per cubic meter, the demand per capita is reduced by nearly 40% than the demand in Puerto Ayora.

The total domestic per capita demand, including all types of sources, is significantly higher in Bellavista (253 lpcpd). This is due to a high consumption from rainwater and water purchased from trucks. Based on these findings, it can be concluded that in general there is no real scarcity of the water resources within the island (190 lpcpd in average is consumed per inhabitant in Santa Cruz) but instead there is a deficient management of water supply and demand.

Fixed tariff structures should be abolished in Puerto Ayora and further reviewed in Bellavista. Revised tariffs will help to increase environmental awareness amongst the population and therefore reduce the demand significantly. Also, the municipality will have higher revenues and the financial means to improve the service. In addition, the installation of water meters needs to be considered in Puerto Ayora, in order to help controlling water losses within premises. This would allow more accurate figures of water demand for different categories and consequently a solid base to change policies regarding WDM. However, numerous factors allow the inefficiency of the service to continue, blocking its improvement. As a result, authorities do not dare to increase tariffs or change the structures because of a possible reaction from local population.

Even though local population affirms to be willing to pay more on their monthly water bills, this is subjected to a better service and potable water. Therefore, water meters must be installed for all connections, in order to increase revenues and lower demand. Simply increasing current fixed tariffs would not be sufficient to cover the deficit of the municipality. Therefore a linear tariff structure or an IBT would be a better option. An IBT is preferable as it promotes WDM and major consumers would pay significantly more. Furthermore, new policies regarding spilling of water when filling tanks must be seriously considered, and penalties may be needed. With a systematic and regular control of this problem, the overall water losses will be lower and there would be additional volume of water for more hours of service.

Finally, it is necessary to determine the exact number of tourist accommodations in Santa Cruz, in order to make more accurate demand estimates regarding the hotel sector. The Ministry of Tourism needs to monitor systematically the establishment of new tourist facilities and enhance the legalisation of the current ones. Also, the promotion of eco-tourism (attracting visitors who understand environmental threats), is needed to control current massive tourism. Unless sustainable measures are developed on time, the Galápagos Islands may lose all their natural attraction and endanger unique species.

"Human nature is like water. It takes the shape of its container."
— Wallace Stevens

5

ASSESSMENT OF DOMESTIC CONSUMPTION IN PUERTO AYORA INTERMITTENT SUPPLY NETWORK

Intermittent water distribution systems are stark reality in developing countries. Puerto Ayora, also has this type of supply, in need of preserving the scarce water resources. However, the wide range of specific consumptions reported in recent studies, well above those in 24/7 supply situations, questions the idea of lack of water, and actually reveals a lack of proper management of the water supply system. In this particular study, 18 water meters were installed, and a water-appliance diary was carried out in 15 randomly selected households in Puerto Ayora. The aim was to analyse the domestic water demand and diurnal patterns in an attempt to verify the consumption figures from the previous chapter, and learn more about such a broad range of domestic demands in predominantly water scarce area. This chapter also elaborates on the correlation of consumption regarding the zones and schedules of distribution conducted by the municipality. The conclusions point no influence of the specific intermittency pattern on the specific water demand, but rather a wide range of different lifestyles, presence of informal accommodations and excessive wastage.

This chapter is based on:

Reyes, M., Trifunovic, N., Sharma, S., and Kennedy, M. (2017). Assessment of domestic consumption in intermittent water supply network: Case study of Puerto Ayora (Galápagos Islands). Accepted manuscript in *Journal of Water Supply, Research and Technology (AQUA)*.

5.1 Introduction:

Water is considered one of the most sensitive issues under the arid and semi-arid climatic conditions, due to their limited water resources (Singh and Turkiya 2013). Several urban settlements located in these regions have been experiencing water deficits, showing no improvements. Because of the population growth trends, as well as the associated water demand growths, the condition is expected to worsen. Therefore, rationing the distribution of water seems to be the only strategy.

Even though water supply infrastructures are designed for a continuous and unlimited supply, this hardly takes place, especially in the developing countries (Vairavamoorthy et al. 2008). Many cities in the developing world have an intermittent water supply system. Therefore, the water is distributed some hours per day and, in some extreme cases, per week. For example, 91% of their water supply systems in South-east Asia are intermittent, as well as in nearly all Indian cities. According to Vairavamoorthy and Elango (2002), this intermittency results often in insufficient pressure in the network, as well as in inequity and limitation in the distribution of available water.

Due to the population and economy growth trends, water demand has exceeded water supply capacity in many places around the world. Unfortunately, the lack of relevant data to analyse this phenomenon is common in these areas. Water demand needs to be accurately assessed in order to ensure the resource for the coming years. Accurate demand measurements are therefore essential for water supply companies. With this information, demand patterns and diurnal peak factors can be calculated more accurately, providing useful inputs for design of networks, reservoirs and pumping stations (Trifunovic 2006).

Furthermore, accurate demand quantification will also result in the calculation of the real range of consumption, making the basis for optimal water tariff structure. According to Arbués et al. (2003), the price of water is the main tool that helps to control the demand. Moreover, Rogers et al. (2002), stated that low water prices cause consumers to forget important economic and environmental issues. In addition, according to OECD (2009), any measure which lowers the water tariffs, rather than encouraging the reduction of water or improving the efficient use of the resource, contributes to potential higher volumes of inputs, outputs, as well as pollution of water. Therefore, an optimal water tariff is considered a step towards full-cost pricing of environmentally harmful activities and may lead to the development of policies which internalize social and environmental costs (Dinar 1998).

In addition, water metering, as well as pricing policy, can be used within residential consumers to manage the quantity of distributed water (Jansen and Schulz 2006), which will help to distribute the water in more efficient and equitable way. Moreover, it is imperative to understand the habits of the population regarding water usage, in order to help authorities develop specific policies addressing the preservation of the resource, as well as consumers' satisfaction. Therefore, the understanding of the user's behaviour within households is also useful for future estimations (Fidar et al. 2010). Based on this, a water appliance analysis will be helpful to develop demand management measures through the installation of water saving devices. Consequently, the water demand quantification will aid to develop specific programs with the aim to reduce demand.

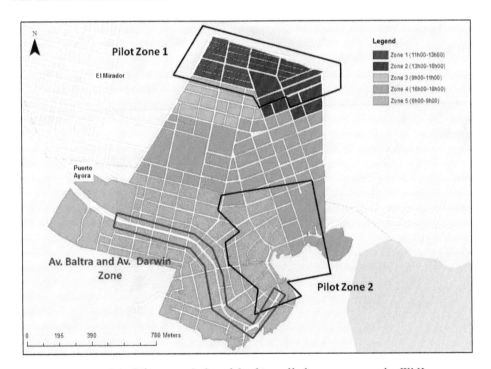

Figure 5.1- Pilot zone 1, 2 and 3 of installed water meters by WMI.

Water demand patterns may vary among countries, cities, villages and different urban settlements and depend on several economic, cultural, climatic, availability and accessibility factors (Singh and Turkiya 2013). Moreover, the determination of the demand patterns in the domestic sector, may improve significantly the effectiveness of the WDM approach (Beal et al. 2011). This type of knowledge will also yield more accurate forecasts, and will contribute to the reduction of future water demand growth. Even though many urban areas are facing extreme stress regarding water, the expansion of their water supply and distribution facilities will be still required. Therefore, in order to ensure a reliable infrastructure expansion strategy, accurate and reliable estimations of water consumption and use are required, especially while assessing the peak demands (Bougadis *et al.* 2005)

The water supply system in Puerto Ayora is not an exception regarding the lack of a continuous water supply system due to the scarcity of the resource. Because of this intermittency, caused mainly by the lack of proper management and by sensitive political issues (Reyes *et al.* 2016), the local population perceive the system as unreliable. The previous chapter suggested the average municipal water demand ranges from 40 lpcpd to 380 lpcpd. The possible subjectivity and uncertainty of some of the responses in the surveys from Chapter 4, pointing the extreme ranges of consumption, questions the real effectiveness of the intermittent mode of supply in Puerto Ayora, where the estimates of the domestic consumption ranges from 163 to 177 lpcpd ± 60 lpcpd.

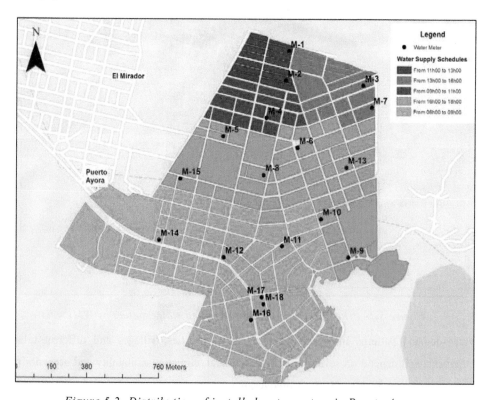

Figure 5.2- Distribution of installed water meters in Puerto Ayora.

Another research carried out by d'Ozouville (2009), estimated the consumption based on the installation of some water meters. This study suggested the specific demand to be as high as approximately 1500 lpcpd, suggesting informal (not registered) accommodations. Furthermore, the same study calculated specific demand based the number of inhabitants and the total supply records from the Municipality of Santa Cruz as approximately 500 lpcpd.

On the other hand, WMI, a private organization financially supported by the German Cooperation G.I.Z, installed approximately 300 water meters in period 2013 to 2015, in three different pilot zones. Pilot Zone 1 (PZ 1), located in the northern part of the town included 115 installed water meters. The readings took place in period from August 2013 to June 2015. This zone was chosen by the municipality due to the prevalence of domestic premises. Furthermore, 140 water meters were installed in Pilot Zone 2 (PZ 2), located in the south-western part of the town. The readings in this zone correspond to the period from February 2014 to June 2015. Lastly, Pilot Zone 3 (PZ 3) was designed to cover the two main avenues of the town where most tourist facilities are located (Av. Baltra and Av. Charles Darwin). The readings in this zone were taken from 54 domestic water meters in period from September 2013 to June 2015. All of these water meters were placed in the distribution network before the individual storage facilities, accounting also for spillage and wastage. The results from these readings indicate average specific demand ranging from 156 lpcpd to 568 lpcpd, for the different pilot zones and the different years. Figure 5.1 shows the different pilot zones where the water meters were installed, as well as the zones and the schedules of intermittent supply according to the municipality.

Figure 5.3-Picture of hand counter installed on toilets

This chapter aims to verify the findings of the survey conducted on water demand estimation presented in Chapter 4, as well as to establish domestic demand patterns using more accurate measurements. Also, it analyses the lack of equity in attempt to find a correlation between the schedules of intermittent distribution applied by the municipality, and the water consumption. The preliminary findings showing extreme high figures of consumption raise the question whether the intermittency is really necessary. Furthermore, the research tries to find any influence of the network's pressure on the household consumption. For that purpose, a small household water-appliances analysis was done with the aim of understanding the end uses of water in the domestic category. Eventually, the aim was to explain the main causes and find justification for use of intermittent supply in Puerto Ayora.

5.2 Research methodology

To get more insight and verify the data obtained in previous researches, a fieldwork was carried out from June to August 2015. In collaboration with the Municipality, 18 water meters (Flodis-single jet turbine device) were installed in private premises based on their willingness to cooperate and the ease of accessibility to install the meters. Figure 5.2 shows the locations of the installed water meters. The water meters were installed on the pipe located after the individual storage facility, therefore not accounting for spillage of tanks or cisterns. The devices were kept for one month (30 days) and then dismantled (for they were borrowed for the specific purpose of this research by the Municipality of Santa Cruz). An assistant was furthermore hired to record hourly readings from 6 am until 8 pm and then the following day at 6 am, to observe cumulative demand overnight. The readings were taken during two working days and one weekend day, to compare and register possible change in habits with respect to water use.

Indoors, a domestic water appliance analysis was carried out at 15 other premises, each representing one neighbourhood. These were chosen based on the willingness to fill a water-appliances diary for one week. The tenants were asked to register the duration of use of each appliance (toilet, bathroom basin, kitchen basin and shower), and the date. Also, hand counters were installed on toilets as shown in Figure 5.3, which were checked at the end of the registration period. The consumption was calculated based on the time registered and the flow of each appliance, which was previously measured. For the toilets, the size of the water tanks was previously calculated as well.

Finally, the information from the water cadastres belonging to the pilot zones was thoroughly assessed for each zone. The results were also analysed with the aim of finding some relation between schedules of distribution and wide ranges of consumption.

Table 5.1-Consumption registered in two working days and a weekend day of 18 installed water meters.

Water Meter	Consumption working day 1 (lpcpd)	Consumption working day 2 (lpcpd)	Consumption weekend (lpcpd)	Average Specific Demand* (lpcpd)	Monthly Specific Demand** (lpcpd)
M1	151	117	208	153	158
M2	101	82	94	91	150
M3	132	78	143	115	140
M4	48	187	216	144	172
M5	257	169	90	176	172
M6	176	114	74	124	288
M7	62	25	76	52	52
M8	247	117	116	162	223
M9	238	238	269	244	275
M10	370	215	97	235	410
M11	140	175	241	179	250
M12	105	71	129	99	86
M13	56	90	80	74	96
M14	80	151	116	114	126
M15	90	54	80	73	71
M16	79	147	51	94	127
M17	78	46	90	69	80
M18	113	50	52	72	75
AVERAGE	140	118	123	126	164
Standard Deviation	±87	±62	±66	±56	±94

*This value refers to the average of specific demand of the three days of measurement and ** refers to the average specific demand based on the monthly measurement taken when the water meters were uninstalled after 30 days and considering the number of inhabitants.

5.3 Results and discussion

The results regarding the domestic water demand in Puerto Ayora are presented divided into different sections.

5.3.1 Results based on installed water meters

5.3.1.1 Water consumption

Table 5.1 shows the values obtained from 18 water meters over a period of one month. The average specific demand refers to the weighted average of the readings corresponding to the three days (Equation 1), and the monthly specific demand refers to the per capita average from the thirty days the meter was installed.

$$Weekly\ specific\ demand = \left(\frac{C_{w1}+C_{w2}}{2}\right) * 0.714 + (C_{wk} * 0.286) \qquad \text{Equation 1}$$

In Equation 1:

C_{w1} = refers to consumption on the working day 1

C_{w2} = refers to consumption on the working day 2

C_{wk} = refers to the consumption in the selected day of the weekend

As can be observed, the average specific demands for the measured week, as well as the specific demands for the whole month, vary significantly. Moreover, most of the average monthly demands are higher than the average weekly demands. Furthermore, the results show surprisingly high figures (up to 410 lpcpd), as well as low demands (minimum of 46 lpcpd). For a deeper analysis, the measurements were further assessed from the perspective of the water meter location in a particular distribution zone and its specific schedule of supply, as well as by anticipating the standard of living of the premises that were metered. Figure 5.4 shows the average per capita demand results based on the location in Puerto Ayora.

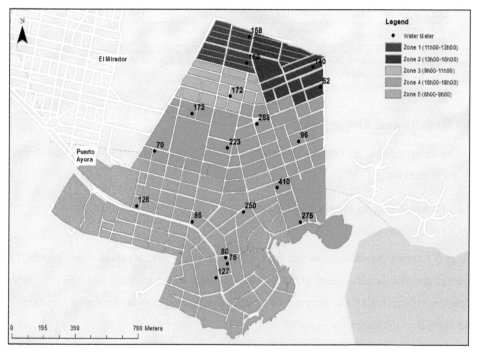

Figure 5.4-Map of Puerto Ayora with water municipal supply distribution zones and schedules, and average specific demand from installed water meters (lpcpd).

Comparisons could not result in correlations between the demands and a specific zone i.e. schedule of intermittency. At first, the possible influence of the network pressure due to the location of the households was considered. Also, it was thought that the households located the furthest from the sources of supply, could experience lower network pressures and consequently the lower consumptions. However, these hypotheses could not be verified because of the random distribution of demands, regardless the location. Remarkably, the meter M10 with the highest observed specific demand, is located in the zone which according to the Municipality has a supply for only two hours per day. Alternatively, these figures may be explained by significant difference in lifestyles and/or volume of individual storage, as well as more negligence in some households than in others, and/or an indication of some sort of informal/illegal tourist business. For example, after the analysis, it was found that one of the households had a small laundry business, explaining the high consumption registered. Another household with high consumption was interviewed, which revealed the negligence by the cleaning/cooking lady who regularly left the faucets open while doing different cleaning activities, as well as the gardening. Furthermore, the inspection of water appliances in some houses, such as in toilets, revealed water losses.

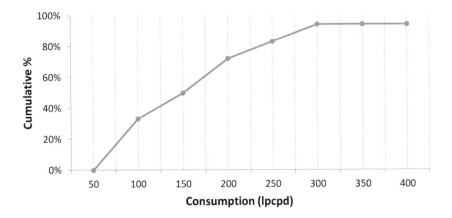

Figure 5.5-Cumulative percentage of consumption from 18 installed water meters

On the other hand, low-demand households were also interviewed in order to find the reasons of the low use. In most cases, it has been observed that those families were spending only part of their time in their homes, otherwise spending their time in the premises of their relatives'. Also, in other cases, some members of the family were on holidays.

Table 5.2-Consumption classification of metered households

Description (lpcpd)	Consumption classification	No. of households	Percentage (%)	Average consumption
< than 100	Low	6	33	76
from 100 to 180	Medium	7	39	149
> than 180	High	5	28	289

5.3.1.2 Consumption classification

Using the average specific demands, a consumption classification was also done.

Table 5.2 shows the demand classification criteria for the metered households, eventually making three groups: low, medium and high demand. In addition, Figure 5.5 shows the cumulative percentage of the specific demand of the households.

Table 5.2 and Figure 5.5 show that on average, most of the population consume more than 100 lpcpd, suggesting that most of the assessed inhabitants have sufficient access to water for common household activites, measured by international standards. Moreover, the cumulative graph shows that 50% of the time, consumption of the metered families is less than 150 lpcpd i.e. excessively, which contradicts the common perception within the population that water availability is restricted. This perception may result from the inflicted intermittency but seemingly does not relate to the quantity of water some households consume.

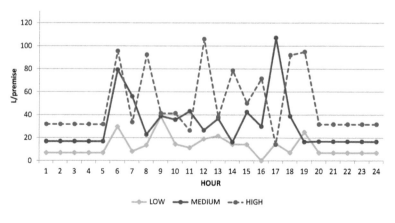

(a)

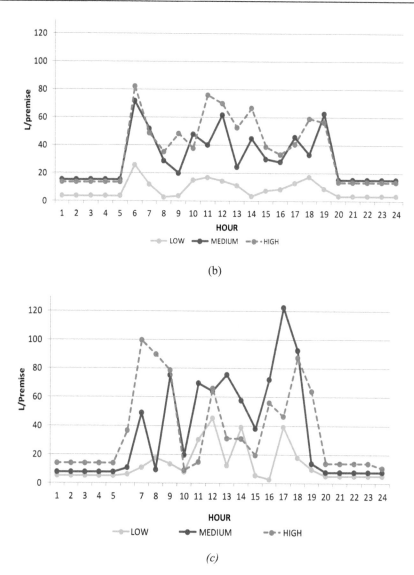

(b)

(c)

Figure 5.6-Average Daily demand patterns for different consumption categories drawn from 18 installed water meters in Puerto Ayora in a) working day 1, (b) working day 2 and c) weekend day

5.3.1.3 Water demand patterns

Daily demand patterns were created from the hourly measurements taken during two weekdays and one weekend day. Figure 5.6 shows the demand patterns based on average consumption for low, medium and high consumers in the three chosen days. Hourly measurements were taken from 6 am to 8 pm each day and then again at 6 am of the following day. An average for the night hours was calculated. Therefore, the night period is represented by a straight line. The schedule of intermittent supply reported by the Municipality does not specify any supply between 6 pm and 6 am, suggesting that the use of water in this period is, exclusively, from the individual storage.

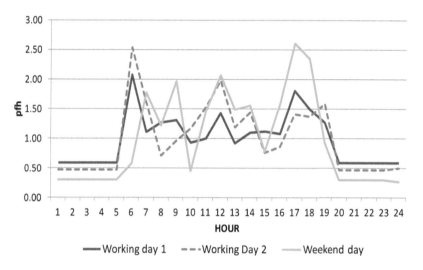

Figure 5.7-Hourly peak factors for three working days from 18 installed water meters

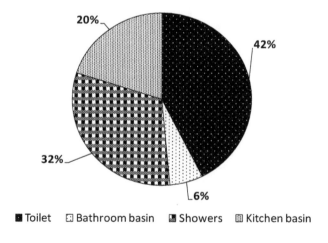

Figure 5.8-Percentages of water use distribution between four selected water appliances.

Figure 5.6 shows that the three categories of consumers listed in

Table 5.2 have similar habits in water use, with three peaks visible during the working days. These peaks are more often in the morning between 6 and 7 am, at midday, and in the evening between 5 and 8 pm. The peak at midday is explained by the common practice of local population to have lunch break at home, due to short distances of their work (small town). These differ from domestic water demand patterns in many other urban settlements with two characteristic peaks (in the morning and evening only).

Furthermore, Figure 5.7 shows the variation in peak factors between different consumption days. The highest peak was observed in the weekend (Saturday), due to the fact that cleaning, laundry and gardening are usually performed during this day. Also, the lower peak was observed in the weekend, portraying the variation in the schedules of the household habits; since it is a non-working day, families tend to leave the premises. Furthermore, in the two weekdays (Tuesday and Thursday), the three peaks (morning, midday and night) are more evident.

5.3.2 Water appliances household analysis

This analysis shows the results of 15 households (one per neighbourhood) who kept diaries over a week, regarding their own water use. Based on the commonly available water appliances, the water use was distributed between the toilet, bathroom basin, kitchen basin and shower, as shown in Figure 5.8.

Not surprisingly, the toilets and showers are the appliances that consume the most of the domestic water. This complies with many studies done worldwide, where toilets are the appliances that consume the most, followed by showers (EPA, 2016). Other water appliances were not included since the selected premises did not have gardens or any outdoor activity using water. Table 5.3 shows the average quantities per appliance, considering the number of inhabitants per household.

Table 5.3-Water use of household appliances per person

Water appliance	Average demand (l/per/day)	Standard Deviation
Toilet	28	±13
Toilet sink	7	±6
Shower	30	±13
Kitchen Sink	25	±29
TOTAL	**90**	**±11**

5.3.3 Comparison of the demand from water meters and water-appliance analysis

Knowing the locations in both cases, the consumptions observed on the 18 water meters were compared with the quantities registered in the 15 water diaries. This relation was drawn based on relative proximity of the households. Yet, the correlation found between the specific demand and the location was extremely low ($R^2 = 0.06$). Figure 5.9 shows the map with the average specific demand from water meters and from the water-appliances diary.

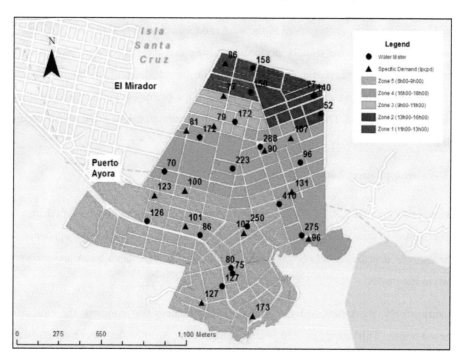

Figure 5.9-Map with average specific demand from installed water meters and water-appliance analysis (lpcpd).

This reconfirms the conclusion that the zones of distribution, as well as the schedules, do not influence the specific domestic water demand. The different specific demands are likely due to different lifestyles and/or the presence of informal (possibly also illegal) tourist business where population numbers might be different, showing the higher specific demands. Also. May be due to negligence of the households.

5.3.4 Analysis of water demand in Pilot Zones

The information from the water cadastres of the municipality were thoroughly organized and analysed. Table 5.4 summarizes the water consumption of the three pilot zones which shows that the average specific demand differs between years and pilot zones. PZ 1 and PZ 2 have similar averages of specific demand, as well as standard deviations, showing a wide range of consumption for different households. However, this range of consumption is significantly wider when compared to the results obtained from the surveys in Chapter 4. The highest average values of specific demand are in PZ 3. However, most of the tourist facilities are also located in this zone, meaning that it could be a mixture of different consumption categories.

Table 5.4-Summary of water demand for three metered pilot zones in Puerto Ayora for 2013, 2014 and 2015

Pilot Zone	Year	Total average consumption per month (m3)	Average consumption per premise (m3)	Specific Demand (lpcpd)	Standard Deviation (lpcpd)
PZ 1	2013	2,689	23	156	±19
	2014	1,039	29	191	±39
	2015	3,449	30	200	±36
	Average	**2,393**	**27**	**182**	**±31**
PZ 2	2013	-	-	-	-
	2014	2,175	24	158	±37
	2015	4,877	35	232	±93
	Average	**3,526**	**26**	**195**	**±80**
PZ 3	2013	2,197	33	217	±32
	2014	4,041	75	499	±88
	2015	4,599	85	568	±92
	Average	**3,612**	**64**	**428**	**±70**

Source: Water cadastres from the Municipality of Santa Cruz

In order to visualize graphically the broad ranges of consumption for the three pilot zones, Figure 5.10(a), (b) and (c) show the maximum, minimum and average specific water demand per month, assuming 5 inhabitants per household.

As observed in Figure 5.10, per capita demands differ significantly between each pilot zone and each household. There are households consuming as high as 1500 lpcpd and others as low as 50 lpcpd in PZ 1 and PZ 2. Evidently, the highest values indicate some sort of informal tourist accommodation or excessive wastage in form of spilling elevated tanks or cisterns. According to a survey done by WMI (2014), some of the metered premises had informal tourist accommodation businesses. Moreover, on PZ 3, there are premises consuming as high as 4500 lpcpd, implying that these "households" are actually not "domestic". Furthermore, most of the peaks observed are in the months of March, April, May and June, which correspond to the warmest months of the year. Also, the lowest consumptions are observed in August, September and October, which are the coldest months of the year, suggesting some influence regarding the season. However, the temperature variations in the Galapaos Islands are not abrupt as in other tourist islands.

Furthermore, these readings were compared with the previous results from the surveys. Since the pilot zones shown in Figure 5.4 consists of several neighbourhoods in Puerto Ayora, the results from the surveys belonging to those specific neighbourhoods were evaluated and then compared with the readings from the pilot zones as shown in Table 5.5. PZ 3 was not included because it encompasses only parts of several neighbourhoods.

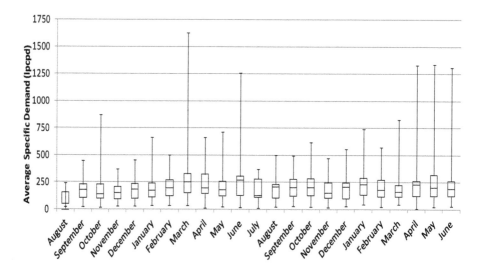

(a)

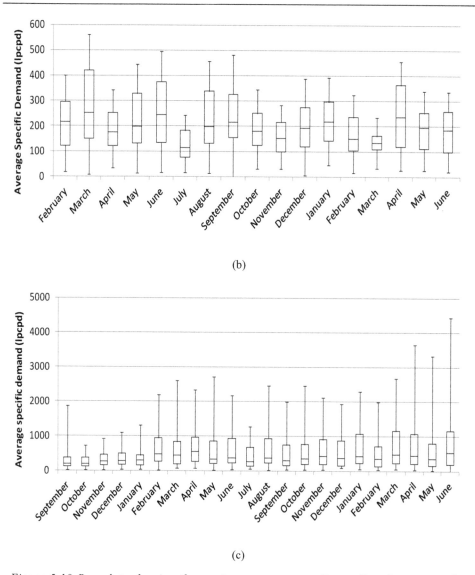

(b)

(c)

Figure 5.10-Box-plots showing the maximum, upper quartile, median, lower quartile and the minimum values of specific demand on (a) Pilot Zone 1, (b) Pilot Zone 2 and (c) Pilot Zone 3

Table 5.5 shows the average specific demand for the neighbourhoods that form PZ 1 and PZ 2, which can also be considered as high, but not as the results observed on Table 5.4Table 5.5 (PZ 1=182 ±31 lpcpd and PZ 2 = 195±80 lpcpd). This is related to the fact that the quantities from the surveys may have been underestimated by the respondents, since the specific demand was calculated based on the storage tank volume and number of times to be filled per week.

Nevertheless, from the standard deviations observed, it is also evident that there is a wide range of consumption among the different users.

Table 5.5-Water consumption for different neighbourhoods based on the survey

Zone	Neighbourhoods	Specific Demand (lpcpd)	Std. Dev. (lpcpd)
	La Cascada	143	±47
Pilot Zone 1	Las Orquídeas	133	±83
	Escalescia	152	±95
	Average Pilot Zone 1	*143*	*±75*
	Pelikan Bay	148	±66
Pilot Zone 2	Acacias	122	±39
	Average Pilot Zone 2	*135*	*±52*

5.3.4.1 Consumption classification on the pilot zones

Based on the information from the water cadastres, the consumers were also classified as low, medium and high, depending on their average specific demand for the entire period each pilot zone was metered. Table 5.6 shows the calculated percentages for each type of consumer.

Table 5.6-Consumption classification for three pilot zones

Description (lpcpd)	Consumption classification	No. of households	Percentage (%)	Average consumption (lpcpd)
		Pilot Zone 1		
< than 100	Low	15	14%	77
from 100 to 180	Medium	29	28%	144
> than 180	High	61	58%	262
		Pilot Zone 2		
< than 100	Low	17	13%	78
from 100 to 180	Medium	43	33%	148
> than 180	High	69	53%	293
		Pilot Zone 3		
< than 100	Low	7	13%	75
from 100 to 180	Medium	6	11%	164
> than 180	High	40	75%	637

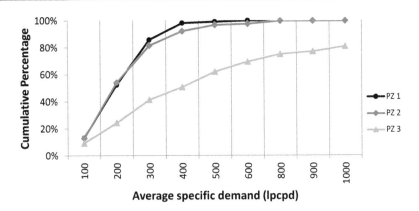

Figure 5.11-Cumulative percentage of average specific water demand for Pilot Zone 1, Pilot Zone 2 and Pilot Zone 3

The analyses of the variation of consumption among different consumers was done by a cumulative graph which helps to observe the percentage of the population which consumes below a certain per capita demand figure. Figure 5.11 shows the cumulative graphs for PZ 1, PZ 2 and PZ 3. As observed in Figure 5.11, the demand in various households is quite high. Once again, these results contradict the perception in a place where water resources are scarce and there is lack of water. By analysing these ranges of consumption from the three different pilot zones of the town, it can be affirmed that there is also a lack of equity in the distribution of water.

5.3 Analysis on scarcity of water in Puerto Ayora

Table 5.7 shows the comparison of the average water supplied per day obtained from the municipal records, against the estimated average water demand from different scenarios (three pilot zones and 18 installed water meters). Based on Chapter 4, where NRW was estimated as ±35%, the leakage figure was assumed as 50% of NRW. Since leakage constitutes the majority of NRW, the assumed figure of 17.5% for leakage is the lowest possible (UNICEF 2000, Chowdhury *et al.* 2002).

By assessing these figures, it can be observed that for most of the scenarios, there is sufficient water to satisfy several average specific demands. Only for the case of PZ 3 scenario (2014 and 2015), the average amount of supplied water would not suffice. However, these averages of specific demand are extremely high and very unlikely to be the representative ones for domestic households. On the other hand, with the specific demand figures of the other scenarios, there seems to be enough water, even though these figures are also high for the domestic category.

This also indicates, one more time, that the perception of limitation of water is not entirely accurate. Furthermore, if the number of unreported tourists in private accommodations is high, it could lower the extreme values of domestic specific demand significantly. If the supplied average is observed, the figure from 2014 increased compared to 2013 by ±1,000 m³/day. Since 2014, one of the pumps has been operating the 24 hours per day (according to the municipality), while the other one is still operating 12 hours per day; in previous years both pumps were on average extracting water only 12 hours per day. Further increase in extraction capacity is questionable due to the extra pressure on the basaltic aquifer, and possible increase in salinity of the resource. Also, because of the fixed tariffs and storage tanks, if more water is available, then more water is most likely to be consumed or wasted (some as spillage of tanks). This seems to be the case for the latest years (2014 to 2016), since the amount of water extracted has increased by 50%, but the total supply hours have not increased at all.

The discussion of the results points to an evident lack of management regarding the water distribution system. With the figures obtained for demand and supply, the quantity of water does not seem to be the key issue, but it is more the quality. As observed from specific demand figures, despite the fact that some premises are registered as domestic, they seem to have some sort of informal accommodation or tourist business, especially in PZ 3. This refers also to the lack of cross-checking of information among institutions. As mentioned in previous chapters, the accurate registration of tourist premises, as well as the information about number of tourists is essential to get real figures about the domestic (and tourist) specific water demand.

Moreover, there is evidence that some households consume extremely high quantities of water. This hints that there is enough water, for at least more hours of distribution than the current ones. It can also be inferred that the storage systems may not be aiding the intermittency, as originally considered, but complicating the situation. This also proves that there is a lack of equity in the distribution of the water, which can be attributed directly to the storage facilities. The bigger the storage facilities, the less the availability of water for the rest of the consumers. Therefore, the individual storage does not necessarily seem to be the best alternative regarding equity within the local population and solving the scarcity issue of the town.

5.4 Conclusions

This chapter addressed the issue of domestic water demand in Puerto Ayora (Santa Cruz Island). For critical analyses of specific domestic water demand, two procedures were used for its estimation. The demand was registered from 18 installed water meters. Also, a water-appliances analysis was conducted at 15 premises, and the meter readings from three pilot zones made available by the Municipality of Santa Cruz were analysed. The average specific demand from the different scenarios provided the same conclusion for all three sources of information: there is a wide range of consumption in Puerto Ayora (averages ranging from 156 lpcpd to 568 lpcpd). An attempt was made to find correlations between the observed demands and supply zones/intermittency schedules determined by the municipality; however, none was found.

The wide range of per capita consumption can be most likely attributed to the different habits within the local population. Also, it can be explained by the lack of awareness and consequent wastage of water within premises, which could be a direct consequence of a fixed tariff structure. Spilling of elevated tanks has been identified as a major problem and it can be supported by the figures in this chapter. Furthermore, the high averages of specific demands also indicate informal accommodations or any other type of business, which may be increasing drastically the consumption and/or involving more (tourist) users.

The reported range of consumption are surprising for an island where water resources are perceived as scarce. Thus, the figures clearly portray that currently there is sufficient brackish water available. Finally, the results obtained also suggest that the intermittency could be reduced if the distribution system would be better managed and WDM practices would be taken into account. For instance, if toilets were to be replaced with more efficient ones (current ones have tanks of approximately 12 litters), it is estimated that the domestic demand could be reduced by approximately 25%. This would also contribute to the improvement of the supply system, allowing more hours of service.

This chapter has shown that the hours of intermittency need to be re-evaluated or even completely eliminated. Storage tanks are not necessarily helping the local population as has been thought over the last two decades, but has contributed to the limitation on the performance of the supply system. This also calls for a change on tariff structures, creating awareness within population. Finally, the leakage needs to be further assessed, since this is an important figure when calculating the amount of total actual supply to consumers. Previous researches have given an idea, but this needs to be confirmed.

Source of info	2013					2014					2015				
	Average specific demand 2013 (lpcpd)	Total demand 2013 (m3/day)	Total Average Supply 2013 (m3/day)	Leakage* 2013 (m3/day)	Remaining (m3/day)	Average specific demand 2014 (lpcpd)	Total demand 2014 (m3/day)	Total Average Supply 2014 (m3/day)	Leakage* (m3/day)	Remaining (m3/day)	Average specific demand 2015 (lpcpd)	Total demand 2015 (m3/day)	Total Average Supply 2015 (m3/day)	Leakage* (m3/day)	Remaining (m3/day)
Pilot Zone 1	172	2,235	3,224	403	586	208	2,699	4,305	538	1,067	219	2,846	4,303	538	919
Pilot Zone 2	-	-	-	-	-	221	2,875	4,305	538	891	207	2,693	4,303	538	1,072
Pilot Zone 3	217	2,819	3,224	403	2	499	6,483	4,305	538	-2,717	568	7,378	4,303	538	-3,613
Water meters	-	-	-	-	-	-	-	-	-	-	164	2,130	4,303	538	1,635

*Leakage was assumed to be 50% of previously calculated NRW by Reyes et al. (2015)

Table 5.7- Estimation of water remaining based on different scenarios of consumption.

"If you can look into the seeds of time, and say which grain will grow and which will not, speak then unto me."

-William Shakespeare

6

MITIGATION OPTIONS FOR FUTURE WATER SCARCITY USING WATERMET² MODEL

An accurate assessment of the future water supply/demand balance is crucial for capital investment on water infrastructure. This chapter aims to develop and present a methodology for similar cases studies, developing suitable options to solve the water supply issues, based on the recently introduced WaterMet² modelling approach. The methodology was illustrated on the main urban settlement of Puerto Ayora, located on Santa Cruz Island. The intervention strategies include environmentally sustainable options such as rainwater harvesting, greywater recycling and water demand management, and as last option the installation of a desalination plant. These strategies were evaluated under four population growth scenarios (very fast, fast, moderate and slow growth) by using several Key Performance Indicators (KPIs) including water demand, total costs, energy consumption, rainwater delivered and greywater recycled. The results obtained show that by 2045 only a small portion of the future water demand can be covered assuming 'business as usual'. Therefore, desalination seems to be the most viable option to mitigate the lack of water at the end of the planning period considering the very high growth scenarios. However, more sustainable and environment friendly alternatives may be sufficient on the slow growth.

———————

This chapter is based on:

Reyes, M., Bezhadian K., Trifunovic, N., Sharma, S., Kapelan, Z. and Kennedy, M. (2017) Mitigation options for future water scarcity using WaterMet2 model: A case study of Santa Cruz Island (Galápagos). Water. **9** (8): 597.

6.1 Introduction

The current chapter aims to develop a water balance for Puerto Ayora, in order to compare the baseline conditions (business as usual) with a number of possible intervention strategies to meet future demand, under different scenarios of population growth rates. This should enable decision makers to investigate the impact of population growth on the level of water services in Santa Cruz Island. The model considers a 30 year period during which four different population/tourist growth scenarios are analysed using the WaterMet^2 software (Behzadian *et al.* 2014).

First, a brief literature review of water demand forecasting using Urban Water Systems (UWS) models is presented and further details of the WaterMet^2 model. Later, the methodology and assumptions applied to the analysed case study are introduced, followed by the description of the modelling approach. This methodology is then applied to the case study of Puerto Ayora. Subsequently, results are presented and discussed and, at the end, relevant conclusions are drawn.

6.2 Water demand forecasting

Water demand forecasting has been developed and applied over the last decades due to a variety of purposes. Mainly, researchers identify it as useful to understand spatial and temporal patterns of water use in the future. Due to water crisis in many regions around the world (UNESCO 2009), prognosis of water consumption is of vital importance for water supply companies. Furthermore, it has also been useful in optimization and planning of water supply and distribution systems, systems' expansions and for calculating future expenses and revenues (Billings and Jones 2008). Moreover, the prognosis of water needs is extremely important for management of water resources, especially in water scarce regions (Ajbar and Ali 2015).

Several methods are used for the estimation of future (urban) water demand, including extrapolating historic trends, correlating demand with socio-economic variables, or more detailed simulation modelling (Donkor *et al.* 2012). Water demand forecasting can be done within different time frames of the prognosis. For instance, short-term forecasting is useful for operation and management of existing water supply systems within a specific time period, while long-term forecasting is important for system planning, design, and asset management (Davis

2003, Bougadis *et al.* 2005). Even though water demand forecast is performed by water utilities worldwide, it is particularly important in regions of limited natural water resources.

The most traditional way to forecast, so far, has been to estimate future per-capita water consumption, and multiply this by expected future population. Moreover, population estimations could be based on different models such as the assumption of current annual growth trend, simple linear growth, a percent annual increase (exponential growth), or more detailed analyses done by demographers. More detailed models take into account a variety of factors, such as changes in population, water prices, climate, customer behaviour, and new regulations (Davis 2003). Although forecasting is not a new discipline, its application in the water sector for demand estimation has encountered many problems, mainly due to the nature and quality of the available data, the numerous variables that are hypothesized to affect the demand (Soyer and Roberson), and the variety of forecast horizons. Therefore, reliable urban water demand forecasting provides the basis for making operational, tactical, and strategic decisions for water utilities (Billings and Jones 2008) and is critical for several reasons.

6.3 Urban water systems modelling

UWS modelling is currently focusing towards a more holistic procedure, trying to include and model every process separately in the system. This new paradigm has changed the approach to urban water services (Mitchell *et al.* 2007) because it holds water supply, drainage and sanitation as components of a whole and complete integrated water cycle. Therefore, this contributes to the main objective of an urban water supply system to balance out demand with supply (dos Santos and Pereira Filho 2014).

There exist several stochastic models of water demand forecasting, which are based on coefficients of water demand time series and other variables, coming from a probability distribution. Moreover, there are also artificial neural network (ANN) tools, which are related to non-linear processes and can be adjusted. ANN modelling is an alternative to stochastic models, and can be trained and verified with two independent datasets. Water consumption time series are therefore combined with weather variable time series and population types, in order to train the ANN forecast model.

General concept	Features	Weaknesses
1. Simulates water and contaminant fluxes in the urban water cycle (water supply, stormwater, wastewater and groundwater). 2. Produces outputs in terms of indicators: water used, stormwater runoff and wastewater emissions 3. Simulates the effect of implementing alternative approaches to water supply, such as rainwater/stormwater harvesting, greywater and wastewater reuse, water efficiency, and test their effects on the different indicators.	1. Models the implementation of urban water management practices: -At land block scale: efficiency of water use, rain- tanks, on-site infiltration of roof runoff, grey water collection, sub-surface irrigation, on-site wastewater collection, treatment and reuse. -At neighbourhood scale: open space irrigation efficiency, aquifer storage and recovery, stormwater infiltration, stormwater collection, treatment and use and local wastewater collection, treatment and use -At study area/development scale: stormwater collection, treatment and use and wastewater collection, treatment and use.	1. Does not cover the full urban water cycle in detail and makes emphasis on catchment and residential area level modelling (runoff-infiltration-evaporation modelling, groundwater modelling, water balance at the catchment or consumer level) → higher data requirement for catchment and residential scale. 2. Water supply component is only represented by a point of imported water, and is not possible to model the different fluxes in other the centralized water infrastructure components. 3. Does not represent changes in centralized water supply infrastructure / water demand over time, i.e. due to demographic variables, changes in water use, and changes in system configuration and/or capacity. 4. Does not produce outputs/indicators of the non-water flows such as energy use and costs (few output indicators).

Source: (Mitchell and Diaper 2010)

Table 6.1–Urban Volume and Quality (UVQ)

Several models were identified in the literature, which cover all the aspects of the urban water cycle using a holistic approach. These models try to include water supply, stormwater, wastewater and groundwater as separate components which become a whole, integrated system. The idea from all these new models has been to take an integrated approach to represent the UWS, simulating both quantity and quality.

Table 6.1, Table 6.2 and Table 6.3 describe three urban water scoping programs (UVQ, CWB, and UWOT).

The following are the three main types of urban water cycle models (Last 2010):

1)   *Detailed models:* These include the softwares such as Infoworks, SMURF, Hydro Planner, which are characterized by limited scope and large requirements of information.

2)   *Catchment scale models:* This type includes the softwares such as Water StrategyMan, Aquatool, Systems Modelling RioGrande, which are characterized by their wide systems' boundaries. However, they do not provide sufficient analysis of the UWS at the sub-city scale, but at a higher (whole) level.

3)   *Urban water scoping models:* This type includes UWOT, Aquacycle, UVQ, WaterCress, CWB and WaterMet²), which model city scale dynamics, including all important processes within the urban water cycle. However, these processes are modelled in less detail than more focused models. Furthermore, they have features such as sustainability assessment tools and strategic planning.

Even though all of these models have positive characteristics, it was necessary to choose the most suitable one for the forecasting of water demand and supply in the island of Santa Cruz. After sufficient literature review, WaterMet² was found to be the most suitable for the objectives of the research. This model, besides forecasting water demand under several population growth scenarios (and tourism scenarios as well), it also allowed the development of several strategic solutions in order to solve the water supply-demand problem in this island.

General concept	Features	Weaknesses
1. Models tool which assesses the sustainability of water management options using alternative scenarios.	1. Wide range of water management alternatives, including sustainable urban drainage system (SUDS):	1. Does not cover the full urban water cycle with detail, more emphasis on modelling at the catchment and residential area level → Higher data requirement for catchment and residential scale
2. Simulates water quantity and quality in the urban water cycle (water supply, stormwater, wastewater and groundwater), and calculates and produces different types of outputs (life cycle energy use and whole life cost).	-At unit block scale: wastewater store & reuse, rainwater harvesting, borehole, greenroofs, swales and soakaways. -At minicluster scale: wastewater and stormwater store and reuse filterstrips, swales and soakaways. -At large scale: wastewater & stormwater store and reuse, large borehole, detention basin, retention ponds	2. Water supply component is only represented by a single point of imported water with a fixed contaminant concentration, energy and costs. Does not model the changes in the centralized water infrastructure components.
3. Simulates the effect of the implementation of alternative approaches to water service provision and tests the effects on the indicators.	2. Calculation of life cycle energy use (using simplified life cycle inventory) and whole life cost. 3. Better representation and model of the natural systems such as groundwater, river, canals, and lakes and models in more detail than Aquacycle/UVQ. 4. Models in great detail runoff-infiltration-evaporation, groundwater, and water balance at the catchment or consumer level. Also, models more efficiently stormwater management and compares sustainable urban drainage systems options.	3. Does not represent changes in a centralized water supply infrastructure, water demand over time, i.e. due to demographic changes, change in water use, changes in the systems configuration or capacity

Source: (Last 2010)

Table 6.2- City Water Balance

General concepts	Features	Weaknesses
-A decision support tool that simulates the urban water cycle. It allows the selection of technological options at 'development' scale (neighbourhood, village or small town) and assesses their effects in terms of sustainability indicators.	-Linked to a database called the "technology library" (presented in Excel) that contains information on the major characteristics (operational or technical specifications) of both in-house and development scale water system components.	-Confined to residential land-use most detailed level of modelling is household micro-components, followed by the household and the development level (neighbourhood/small town). -Limited water management options.
-The technological options for urban water management that can be modelled are: indoor water efficiency options (water saving appliances), household/centralized rainwater harvesting, household/centralized grey water recycling, and a generic sustainable urban drainage system option.	-MATLAB/ Simulink computes the water mass balance model, processes them into outputs in a sustainability assessment, optimisation and visualisation, which is finally presented in Excel.	-The natural system (soil stores, ground water and surface flow) are outside the system's boundary. -Water supply component is presented by a point of imported water, not able to model the fluxes in the centralized water infrastructure components.
-Indicator outputs include cost, energy, water use, wastewater flows and stormwater flows. -consists of two modes: (a) user-defined assessment includes configuration of an urban water management system and (b) optimisation includes technology selection proposed by genetic algorithms (GA).	-Technology is either selected by the user or proposed by optimization of GA, with quantitative and qualitative sustainability criteria and indicators. The optimisation phase provides alternatives based on the full range of sustainability metrics and decision maker preferences.	-Does not represent changes in centralized water supply infrastructure / water demand over time, i.e. due to demographic changes, change in water use, and changes in system configuration/capacity. -Does not simulate or represent water supply-demand balance (i.e. to show any water supply deficit); the water supply is predefined by the user to meet the demand at specific time in the future/specific scenario.

Table 6.3- Urban Water Optioneering Tool (UWOT)

Source: (Makropoulos et al. 2008, Rozos et al. 2009)

Furthermore, it is characterized by the small amount of input information needed, in comparison to other models presented previously. Data limitation is a significant constraint for the case of the Galápagos Islands.

The following are the justifications for the selection of this WaterMet^2 model:

(i) The modelling scope (in the water supply component) is in separate phases, regarding the point of water withdrawal to water consumption point. Therefore, each phase can be simulated separately, and they do not need to include the other.

(ii) Assesses the current as well as the future water balance on the UWS. None of the other models is considered as a holistic systemic perspective for the analysis of resource flows and their impacts on the future performance of UWS.

(iii) It assesses and forecasts the implementation of centralized and/or decentralized intervention strategies, using output indicators (water flows, costs, energy uses, etc.)

(iv) This model does not focus on the modelling of natural systems, including surface and/or water groundwater, which is more suitable technique for the study of urban drainage systems and groundwater recharge-extraction balances.

(v) It has a holistic approach, which includes strategies for long-term planning. It also does not have excessive data requirements for partial modelling of the system. These two characteristics are addressed through a metabolism-based approach, which refer to the different fluxes and conversion processes related to all water flows, materials and energy in a UWS.

(vi) It has a wide variety of output indicators, which indicate the different fluxes such as water flow, energy flow, greenhouse gas emissions, acidification/eutrophication flow, pipeline material flux, chemical flux, pollutant flux. This offers also a wide choice of (sustainability) indicators that allow assessing the selected intervention strategies.

(vii) It is able to represent the daily, seasonal, and annual (future) dynamics of the water demand, caused by demographic demand patterns changes.

6.4 WaterMet² model

6.4.1 General description of the model

WaterMet² is a quantitative UWS performance model, which was developed with the aim of assessing long-term and strategic performance of the UWS. Also, it allows to plan under different and possible future scenarios (Behzadian *et al.* 2014). This assessment tool was developed as part of the TRUST (Transitions to the Urban Water Services of Tomorrow) Project. TRUST is an integrated research project which deals with research innovations and tools that create a more sustainable water future (TRUST 2014).

WaterMet² model is able to quantify the metabolism-related performance of the integrated UWS. Behzadian *et al.* (2014) explain that the metabolism refers to the flows and conversion processes of all kinds of water, materials and energy fluxes in the UWS. The WaterMet² simulates the mass balance and calculates the principal water-related flows, as well as all other system fluxes (e.g. energy, greenhouse gas emissions, chemicals, pollutants, among others). All of this modelling enables different indicators to explain the current or future (over a pre-defined planning horizon) levels. Some of these indicators include total amount of water supplied, energy used, greenhouse gases emitted, and operation and maintenance costs, among others. It also includes the key components in water supply, water demand, and wastewater schemes. The UWS performance can be simulated over a long period of time with a minimum of daily time steps and it is also able to simulate water consumption points into different parts (Behzadian *et al.* 2014).

The traditional approach for UWS modelling is generally physically-based. It simulates the hydraulic behaviour of an UWS, demanding substantial input data, but it only addresses a part of the entire water system (Behzadian *et al.* 2013). WaterMet² is a "simplified and conceptually-based model which can be used to simulate the overall performance of the entire UWS and to assess the impact of any specific change in a component of a whole system". This makes the wide scale modelling approach faster and also includes holistic and sustainable approaches (Behzadian *et al.* 2014).

The results of the simulations can obtained with different quantitative key performance indicators (KPIs). These are used as metrics for evaluating specific criteria, especially when different scenarios are to be compared. The results may be presented with graphs or tables, and can be distinguished by categories. The KPIs show the effects of external drivers such as urbanisation, population growth, aging infrastructures, etc. to the entire UWS system, allowing the decision makers to assess the impact of implementation of intervention strategies in the UWS (Behzadian *et al.* 2014).

6.4.2 Main concepts of the model

6.4.2.1 Fluxes simulated in the WaterMet^2 model

Several types of water flows are simulated by WaterMet^2.

(1) Clean (potable) water: (treated) water originally supplied from the water sources.

(2) Storm water or rainfall from impervious and pervious areas.

(3) Grey-water or wastewater originated from water already used.

(4) Green water: (treated) rainwater modelled in local area scales.

(5) Recycled (reused) water: return water treated by either centralized or decentralized processes.

(6) Wastewater (black water): used water obtained from toilet and polluted water which needs to be collected and treated.

The water quantities can be modelled in all of the subsystems, while the water quality is only modelled in the wastewater and water recovery subsystems (Behzadian and Kapelan 2015).

Furthermore, other fluxes included in the UWS are: (1) energy flow; (2) greenhouse gas emission (GHG) flow; (3) monetary flow and (4) chemical flux. These flows are calculated in each pre-determined time step, based on the amount of the water conveyed and/or treated, and the amount of the consumed flux per unit of water (volume). Furthermore, the energy (electricity or fossil fuel), GHG emissions, acidification, eutrophication and O&M cost flows are calculated based on the total volume per day of water transported by each conduit in the distribution system (including leakage).

6.4.2.2 Spatial levels and its characteristics

The WaterMet^2 software has four main subsystems, which model all the flows and fluxes as illustrated in Figure 6.1. These four main subsystems are: (1) water supply subsystem, (2) water demand subsystem, (3) wastewater subsystem, and (4) water recovery subsystem (comprising of grey-water recycling (GWR) tank and rainwater harvesting (RWH) tank in centralized or decentralized regimes).

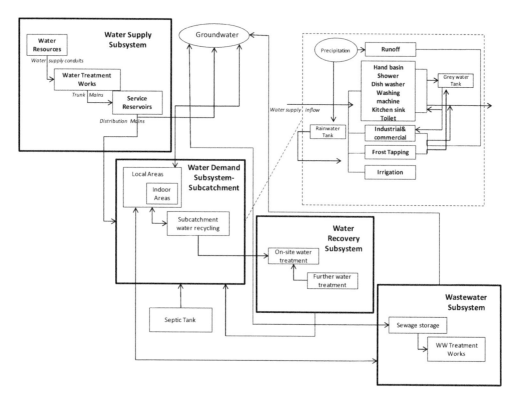

Figure 6.1- Spatial levels and components of WaterMet² model

6.4.2.2 Spatial levels and its characteristics

The WaterMet² software has four main subsystems, which model all the flows and fluxes as illustrated in Figure 6.1. These four main subsystems are: (1) water supply subsystem, (2) water demand subsystem, (3) wastewater subsystem, and (4) water recovery subsystem (comprising of grey-water recycling (GWR) tank and rainwater harvesting (RWH) tank in centralized or decentralized regimes).

The main approach used in the entire system is a mass balance of water. The water sources, as well as the subsystems simulate the boundaries where water is supplied and/or received. The water storage components mean any physical asset which stores water in the different processes. Furthermore, these storages are interrelated with each other through a variety of defined water flows.

This model adopts the different limits of an urban water utility, represented by spatial limits of the UWS. It recognizes four spatial scales that represent the entire UWS: (1) indoor area; (2) local area; (3) sub-catchment area; (4) city area.

1) The indoor area scale is the smallest one of the UWS and represents a single household or premise, either residential, industrial, commercial, public, etc.). It does not include any surroundings such as gardens, public open spaces or any other outdoor area. Therefore, indoor water consumption is defined at this level (such as specific per capita demand).

2) The local area is a group of similar households or premises, including surrounding areas that contain any number of indoor areas of the same type (identical specific water demand). The surrounding area is divided into pervious surfaces, impervious surfaces, and water bodies (such as lakes and rivers). The main operation of a local area is to handle outdoor water demands, rainfall-runoff modelling, as well as local RWH and GWR systems. Irrigation and public water use are also defined at this level.

Table 6.4- Components modelled at different spatial levels in the WaterMet² model

Spatial Level in the WaterMet2 Model		Indoor Area	Local Area	Sub-catchment Area	City Area
Component	Description				
Water Consumption Points	Including Indoor and Outdoor water usages	√	√		
Rainwater Harvesting Tank	Collection and treatment of rainwater from impervious areas for water reuse		√	√	
Grey Water Recycling Tank	Collection and treatment of greywater from water consumption points for water reuse		√	√	
Water Supply Conduits	Conveyance of raw water from water resources to WTWs				√
Trunk Mains	Conveyance of potabe water from WTWs to service reservoirs				√
Distribution Mains	Distribution of potable water from service reservoirs between water consumption				√
Combined/Separate Sewer Systems	Collection of sanitary sewage/storm water runoff from subcatchment area and conveyance to WWTWs				√
WTWs, WWTWs	Treatment of raw water and wastewater				√
Service Reservoirs	Potable water storage prior to distributing between the customers				√
Water Resources	Raw water sources for WTWs				√

Source: (Behzadian *et al.* 2014)

3) The sub-catchment area represents a group of next-to-each-other local areas. This area is defined mainly by topology of the place, as well as the type of wastewater and/or storm water collection schemes. The sub-catchment and local area sizes depend on the size of a city or area

to be modelled, as well as on the level of spatial resolution required, type of intervention modelling, and available data in this level. In this area, larger and/or centralised RWH and GWR systems are modelled.

4) The city area is the top level of the modelling resolution. Several components are defined and modelled only at this level: water resources, water supply conduits (WTCs), water treatment works (WTWs), trunk mains, service reservoirs, distribution mains, sub-catchments, separate/combined sewer systems, waste water treatment works (WWTWs) and receiving waters. The city area can be divided into any number of sub-catchments, which is highly dependable on the available data and the level of interventions to be incorporated.

A list of all the different processes, as well as the components that can be modelled at the different spatial levels of the WaterMet² are summarised in Table 6.4

6.4.3 Water supply subsystem

This model adopts a simplified approach of the water supply subsystem, in which a modelling strategy of 'source to tap' is applied. There are three elements modelled in this subsystem: 1) Three 'storage' components including raw water resources, WTWs, and service reservoirs; 2) Three principal flow 'routes' including water supply conduits, trunk mains and distribution mains; and 3) Sub-catchment representing the point where water is consumed. The hierarchy of water supply subsystem components is shown in Figure 6.2.

6.4.3.1 Water resources components

Three types of resources are recognized in this subsystem: (1) surface water (e.g. lake, dam reservoir, or river); (2) groundwater and (3) seawater desalination. Surface water needs definition of specific storage, defining the storage capacity, initial volume, annual water loss, and time series of inflows. For the surface water reservoirs it is necessary to include individual daily time series of inflow. On the other hand, this model does not consider storage for groundwater and desalination. For this reason, the groundwater resources and desalination characteristics are needed such as: percentage of water losses, electricity consumption, fossil fuel consumption, operational costs, and diverted flows.

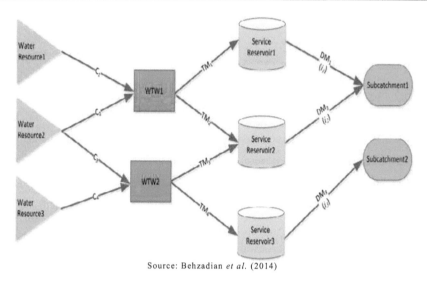

Source: Behzadian *et al.* (2014)

Figure 6.2- Hierarchy of water supply subsystem components in WaterMet²

6.4.3.2 Water supply conduits

The water supply conduits refer to means of conveying raw water from the resources point to the WTWs, and then the treated water is transferred to service reservoirs. The main components are daily treatment capacity, percentage of water loss, electricity and fossil fuel consumption, operational costs, chemical consumption, chemical costs, amount of sludge generated, and diverted flow. The water losses take into account the percentage of treated water which is leakage, as well as water that might be removed by evaporation or infiltration. The water demand and other flows, such as delivered or undelivered flows in each WTW are calculated similar to the water supply conduits, described in the previous numeral.

6.4.3.3 Water Treatment Works (WTWs)

The main components are daily treatment capacity, percentage of water loss, operational costs, chemical consumption and costs, amount of sludge generated, and electricity and fossil fuel consumption. The daily treatment capacity controls the maximum quantity of daily treatment, and the water loss considers the amount of treated water which is removed from the water flow by evaporation or infiltration. The water demand and other water in the WTW are calculated similar to the conduits.

6.4.3.4 Trunk mains

Treated water in WTWs is further conveyed to the service reservoirs by trunk mains. The components are transmission capacity, percentage of leakage, electricity and fossil fuel use, and operational costs. The water flows and other fluxes such as energy and O&M costs are calculated similarly to the fluxes of the water supply conduits.

6.4.3.5 Service reservoirs

The main function of service reservoirs is to store treated (potable) water before the distribution to the different sub-catchments. This are supplied by WTWs through the trunk mains and further distributed through distribution mains. The main components are storage capacity, percentage of annual water losses, initial volume for the first day, operational costs, chemical costs and use.

6.4.3.6 Distribution mains

These are the last components of the water supply subsystem. They convey (potable) water to the different points of consumption. The distribution mains in WaterMet² are characterised by a number of constant parameters over the selected planning horizon, which include: daily water transmission capacity, annual rehabilitation, leakage percentage, energy in form of electricity and/or fossil fuel required for the distribution of a unit volume of water and fixed annual operational costs.

6.4.4 Water demand subsystem

The water demand subsystem includes all the water consumption points, either drinking or non-drinking water in the UWS. Drinking water demand can only be supplied through the water distribution network outlined, while non-drinking water demand can be supplied by other sources (potable or non-potable), such as RHW. This subsystem has three other (lower) spatial levels of the UWS which include indoor area, local area and sub-catchment. Therefore, the software divides the water demand into two main categories of consumption: indoor and outdoor.

Water demand types are defined in the local area level, which include: indoor water, industrial/commercial, irrigation, frost tapping and unregistered public use (defined as m³/day, except indoor demand which is defined by lpcpd). For the latter, an average occupancy per household unit needs to be specified. WaterMet² recognises the variations of water demand in three temporal scales:

1) Annual variations: are defined as the coefficient of annual population growth for the different categories/types of water demand: This is often used for the assessment of different scenarios of population growth in the UWS.

2) Monthly variations: monthly pattern coefficients are specified, used mainly for the purpose of calibrating the model.

3) Daily variations: daily water demand variations by defining two periods of time for each of water demand type, and the correlation of each type of daily water demand to the temperature variations, and by correlating each type of daily water demand to the temperature variations. At this level, contribution of temperature variation is added by the following relation:

$$Cd_{ijk} = 1 + \frac{T_k - T_{ave}}{T_{max} - T_{min}} \; x \; F_{ij}$$
(Equation 7.1)

Where,

Cd_{ijk} = coefficient of daily variations for demand category i, local area j and day k;

F_{ij} = fluctuation factor accelerating the temperature effect for demand *category i* and local area j;

T_k = Temperature for day k in Celsius degrees;

T_{max} and T_{min} = absolute maximum and minimum temperature of the study area, respectively (in Celsius degrees).

It is recommended to define the coefficient F_{ij} between -1.5 and +1.5 in order to avoid generating negative values for coefficient Cd_{ijk}, even though negative coefficients are converted to zero. Finally, the actual water demand for category i, local area j and day k of the year M (Da_{ijkM}) can be calculated as

$$Da_{ijkM} = D_{ij} \; x \; Cd_{ijk} \; x \; Cm_{ij} \; x \; \sum_{m=1}^{M} Ca_{ijm}$$
(Equation 7.2)

Where,

D_{ij}=average water demand for category i, local area j (m³/day);

Cm_{ij}=coefficient of monthly variations for category i and local area j;

Ca_{ijm}=coefficient of annual variations for category i, local area j and year m.

If there are two or more types of water sources available, the WaterMet² provides categories to: (1) rainwater at local area; (2) grey water at local area; (3) rainwater at sub-catchment level; (4) grey water at sub-catchment level; (5) treated grey water from centralised system; (6) (potable) water from distribution mains. The first five options are available only when the user activates and defines the tanks in order to allocate water reuse for the different categories of water demand.

There are two ways to define indoor water demand: by defining per capita indoor water consumption, or by indicating the percentage of contribution per each water appliance, such as hand basin, bath and shower, kitchen sink, dish washer, washing machine and toilet. By specifying this, the user may explore a wider range of intervention strategies related to the water demand management options.

6.4.5 Rainfall, run-off and infiltration modelling

WaterMet² uses a simple rainfall-runoff algorithm, 100% runoff from impermeable surfaces. The rainfall-runoff simulation is modelled only on local area level, considering it the smallest hydrologic unit area. A local area can be divided into two types of surfaces: pervious and impervious areas. The impervious area has no infiltration and only depression storage which transforms all rainfall to runoff. The impervious area is divided into roofs and pavements and roads. By defining two distinct impervious areas it gives the possibility to separate which source can be used for RWH in a local area. Thus, the daily volume of runoff (V) for each local area is calculated as follows:

$$V = C \times h \times (A_{roof} + A_{road} + (1 - i) \times A_{pervious}) \hspace{2cm} \text{(Equation 7.3)}$$

Where,

C=   runoff coefficient between 0 and 1

$h =$ height of daily rainfall and snowmelt (cm)

A_{roof}, A_{road}, $A_{pervious}$= total area of roof, road & pavement and pervious areas in a local area, respectively (m^2);

$i =$ infiltration coefficient for pervious areas between 0 and 1 to account for infiltration through garden, park and open spaces in urban areas.

Furthermore, rainfall-runoff is also modelled as rain and snow, represented by a time series of weather conditions as input data. The data includes precipitation levels and type, snow depth, mean temperature, average wind speed, hours of sunshine, mean relative humidity and vapour pressure on daily basis.

6.4.6 Water resources recovery subsystem

There are two types of water resources recovery subsystems including centralised and decentralised facilities. The centralised water recovery in WaterMet^2 is modelled at a city scale, while decentralised schemes (RWH and GWR) are modelled at sub-catchment and local area levels.

6.4.6.1 Rainwater harvesting scheme (RWH)

RWH scheme can collect runoff from impervious surfaces, such as roof, road or pavements of any local area. Then it can be stored in a pre-specified tank that can be later be used for non-potable water purposes, such as green water. Any of these two inflows can be defined by the user as the source for a local area RWH tank. The use of this RWH tank is also defined by the user to be utilized for toilet flushing, dish washer, hand basin, kitchen sink, shower, washing machine, industry and/or irrigation. The water volume exceeding the capacity of the tank may overflow directly to the sub-catchment RWH tank and/or local area GWR tank. If none of these options are specified, the overflow will be discharged into the sewer system.

6.4.6.2 Grey water recycling (GWR) scheme

GWR scheme collects grey-water and recycles it for some water demand in a local area. The smallest scale of GWR scheme modelled is at local area. The local area GWR scheme can receive two types of inflow: (1) grey water used in local area from dish washer, hand basin, shower, washing machine, industry and/or frost tapping; (2) overflow of local area RWH tank. The outflow at both levels is used for water demand that has to be defined by the user. The overflow of local area GWR tank can be discharged into the sub-catchment GWR tank or the sewer system.

6.4.7 Modelling water resource recovery subsystem

Modelling water resource recovery in WaterMet² requires setting the parameters in two steps:

1) Allocation of water recovery to different water demand categories: refers to the allocation of recycled water from the sources of water recovery (RWH and GWR schemes) for each water demand category (toilet, dish washer, hand basin, kitchen sink, shower, washing machine, industrial and irrigation).

2) Specifying decentralized/ centralized recycling water: The required characteristics include the specifications of the recycling system, sources, sinks of water flow, and place of consumption for stored/treated water.

6.5 Research methodology

In this study, the WaterMet² model was used to forecast the urban water balance for a future 30 year period. First, four different growth scenarios were chosen to predict and assess future water demand based on previous studies. Afterwards, after sufficient literature review, the model was built for Puerto Ayora case study and the baseline condition was analysed, in order to develop relevant strategies that will solve the future water deficit. Based on this, six water supply and demand management alternatives (individual strategies) were analysed in the Puerto Ayora model. The impact of each of these alternatives to meet future water demand was assessed by analysing the percentage of coverage of water demand with supply at the end of the planning horizon (fraction of water demand delivered as Key Performance Indicator-KPI).

Due to the low fractions of water demand delivered at the end of the planning horizon, these individual alternatives were further combined in order to improve the future coverage of water demand with supply, developing five more complex intervention strategies. These new strategies combined also several sustainable options, recurring to desalination as the last option. These analysed strategies were compared using a number of KPIs in order to analyse the impact of the selected growth rates on different aspects of water demand and water supply over the period selected. The KPIs used include the ratio of water delivered to consumers, total costs (i.e. capital and O&M) and total energy use (i.e. direct and embodied). With this indicators, each strategy was assessed in order to find the most sustainable and most optimal for this case study, considering the fragility of the ecosystem, which are addressed in the discussion.

Finally, conclusions were drawn, assessing all the KPI's used and the extension to which each strategy would comply at the end of the planning horizon. This, with the intention to portray to local authorities and stakeholders the limitations and benefits of each strategy included in this study.

6.6 Population and tourist growth scenarios for Puerto Ayora

The population growth scenarios were chosen based on the suggestions made by Mena *et al.* (2013), according to historical growth and governmental planning. This was conducted by deriving relationship between the number of tourists and local residents in Galápagos, representing a demographic model with projections until 2033. In that study, an ordinary least squares (OLS) linear regression was used to model the relationship based on population censuses in 1982, 1990, 2001 and 2010. This resulted in determining the number of residents in each year based on the corresponding number of tourist arrivals. Only land-based tourists were considered (excluding tourist cruise/ships). Three main growth scenarios were developed by Mena *et al.* (2013) as zero (1%), moderate (3.5%) and fast growth (8%) depending mainly on the migration rate since it is the primary demographic parameter due to the strong ties between the growth rate of local residents and tourism growth. More specifically, the growth scenarios developed for this chapter were defined as follows:

Slow growth: The tourist arrivals were 180,000 in 2012, therefore that number is considered to be the average per year. This scenario is suggested and preferred by environmentalists, NGOs and the Galápagos National Park.

Moderate growth: The tourist arrivals maintain an average annual increase of 7,066 visitors (4%). This figure was estimated from the recorded tourism growth in the last 20 years.

Fast growth: The tourist arrivals increase exponentially by an annual rate of 7%. This scenario is preferred by the central government following their objective to increase tourism revenues in the whole country.

Very fast growth: The number of tourist arrivals would be eight times greater than the number of residents at the end of the planning horizon (i.e. in year 2044), suggesting a rate of annual growth of 9%. Also, a scenario is preferred by the central government.

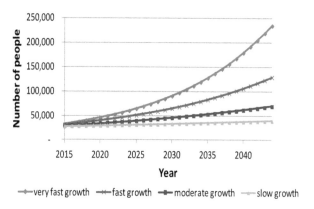

Figure 6.3- Projection of local population growth in Galápagos Islands

A summary of growth rates used in the four scenarios are shown in Table 6.5. The tourism growth scenarios are based on the number of land-based tourist arrivals per year, which includes exclusively the tourists that stayed in hotel or other land-based accommodations per year. The number of sea-based tourist arrivals was not included because the water demand for the sea-based tourists is provided independently by the cruise tourism company and it is assumed that it will not affect the total municipal water demand.

Table 6.5-Annual population and tourist growth scenarios used for water demand forecast

Growth scenario	Local population increase	Tourist visitors increase*
Very Fast	7 %	9 %
Fast	5 %	7 %
Moderate	3 %	4 %
Slow	1 %	1 %

*The historic average growth per year is 7%

Based on the annual increase percentage presented previously, the number of population and tourist arrivals per year were calculated and are shown in Figure 6.3 and Figure 6.4, based on Mena *et al.* (2013). The projection was done for the length of the planning horizon from 2014 to 2044 (30 years). Figure 6.3 shows that the number of local population will increase from 30,000 in 2015 to 40,000 inhabitants in 2045 for the slow growth scenario, and to 240,000 inhabitants for the very fast growth scenario.

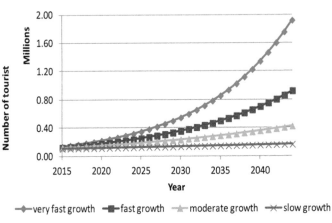

Figure 6.4- Projection of land based-tourist arrivals in the Galápagos Island

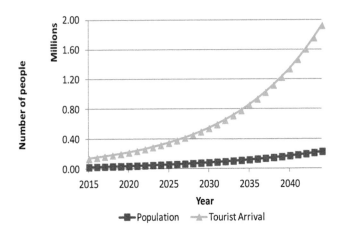

Figure 6.5- Tourist and local population growth in the Galápagos Island (very fast growth scenario)

Furthermore, Figure 6.4 shows that land based tourist arrivals will increase from 135,000 in 2015 to 170,000 tourist arrivals in 2045 for the slow growth scenario, and to 1,900,000 tourist arrivals for the very fast growth scenario. Figure 6.5 shows the comparison between number of tourist arrivals and local residents in Galápagos Island from 2015 to 2045.

Figure 6.5 shows that in the case of the very fast growth scenario, the number of tourist arrivals per year is eight times the number of the local population in 2045. However, the average number of tourists on any given day of the year in 2044 is (only) 6,530 tourists per day (Mena *et al.*

2013). Thus, the high number of tourist arrivals per year, does not necessarily make the tourism-related water demand the major part of the total water demand, but the local population.

6.7 WaterMet² model building

The main input data used to model the UWS of Puerto Ayora in WaterMet² are shown in Figure 6.6 and are divided into three primary categories: 'Water Supply', 'Sub-catchment' and 'Water Resource Recovery'.

The 'Water Supply' specifications of the storage components are storage capacity, initial volume as well as energy, chemicals and cost used per unit volume of water. The second type (i.e. transmission) component of the water supply system is the connecting flow routes for conveying water between storage components in the WaterMet² model. Their general specifications include transmission capacity, leakage, energy and O&M costs per unit volume of transmitted water. The detailed input data of water supply components used in the WaterMet² model for Puerto Ayora is presented in Table 6.6. The theoretical energy consumption was calculated based on the following equation:

$$N = \frac{\rho g Q h_p}{\eta_p}$$

Equation 7.4

Where, ρ is density (kg/m³), g is gravity constant (m/s²), Q refers to the flow (m³/s), h_p refers to the pumping head (m) and η_p is the efficiency of the pump.

The sub-catchment and local area components were used in WaterMet² to define water demand categories and rainfall-runoff characteristics of the model. Puerto Ayora was represented as a single sub-catchment with a single local area. Two water demand categories were defined: (1) 'indoor' water representing domestic water use and (2) 'industrial' representing water demands of restaurants, hotels and laundries. The percentage share of water use for appliances and fittings in domestic water consumption in both cases were assumed to be 7% for hand basin, 20% for kitchen sink, 24% for showers and 49% for toilet flushing, based on the study made on Chapter 5. Furthermore, rainfall-runoff simulation was modelled considering runoff from roofs only (local area level). For the software calculations, roof area proportion and pervious areas proportions were obtained from the literature. A summary of the input data used for modelling the sub-catchment components in WaterMet^2 is given in Table 6.7.

Other input information includes unit costs (of electricity and diesel fuel, water meter installation cost, and inflation rate), climate constants (elevation and geographical location to be used in rainfall-runoff modelling for calculation of evaporation), coefficients for all water demand categories, including percentage of conversion from water to wastewater (assumed here as 95%), percentages of domestic water appliances and their possible energy consumptions, based on personal communications with personnel from the Municipality of Santa Cruz. The historic time series of weather data (e.g. precipitation, temperature and etc.) for the past 30 years were used here in the WaterMet^2 model assuming that the same trend will happen in the future planning horizon. Time series data were obtained from the National Institute of Meteorology and Hydrology of Ecuador (INAMHI). It is relevant to mention that the annual average rainfall over the last years in Puerto Ayora is 380 mm, but in other settlements located higher, such as in Bellavista the average is 1100 mm or even higher were the annual average can reach 2500 mm, having significant higher precipitation rates on the hot "invierno" season, than in the cold "garua" season, characterized by big and strong events of rain. On the other hand, the evapotranspiration average on both seasons is around 400 mm.

The water demand variations of local areas can be defined in WaterMet^2 for different temporal scales (i.e. annually and monthly). The annual variations were analysed under four selected population growth scenarios while the monthly variations were adjusted during the model calibration process, which will be discussed later. The daily variations and temperature influence on these variations was ignored in this step due to the lack of daily consumption registration.

'Water Resource Recovery' refers to RWH and GWR schemes. In this study, rainwater was assumed to be collected only from roof runoff and provide water for toilet flushing, showers, sinks, indoor irrigation and commercial uses. GWR collected from hand basins and showers was allocated only for toilet and indoor irrigation. The associated costs and energy was also considered for treatment and purification of RWH and GWR, since the existing supply system does not have any rainwater and/or grey-water infrastructure; thus, these are considered as new alternatives.

Table 6.6-Input information for water supply component

Water Resources Form		
Component name	**Unit**	**Puerto Ayora**
Type	-	Groundwater
Energy consumption (electricity and fossil fuel)	kWh/m³ (for electricity) L/m³ (for fossil fuel)	0.66 kWh/m³ and 0.3 L/m³
Fixed annual operation costs	EUR/ year* (Cost of elec: 0.17 EUR/kWh and Cost of fuel: 0.22 EUR/L)	219,120
Water Loss	%	Assumed there are no water loss at the point of extraction.
Water Supply Conduits		
Component name	**Unit**	**Puerto Ayora**
Transmission capacity***	m³/day	3,024
Leakage **	%	8
Pumping system	m³	N/A
Energy consumption	kWh/m³ (for electricity) L/m³ (for fossil fuel)	0.66 and 0.5
Fixed annual operation costs	EUR / year* (cost of elec: 0.17 EUR/kWh Cost of fuel: 0.22 EUR/L)	4,980
Service Reservoirs		
Storage Capacity***	m³/day	3,024
Initial volume	m³	1,500
Operational cost	EUR / year*	2,490
Distribution Mains		
Transmission Capacity	m³/day	3,024
Leakage	%	20
Operational Costs	EUR / year*	58,100

*All the costs are based on information from the municipality **Leakage figure of 28% was used and not 35% as previously identified as NRW, due to calibration purposes. *** Transmission components in WaterMet² (e.g. water supply conduits and distribution mains) are defined based on a daily transmission capacity expressed in m³/day while storage components (e.g. service reservoirs) are defined based on storage capacity expressed in m³. The definition for transmission components is for a conceptual model used in WaterMet² and is estimated based on average hydraulic capacity of the components.

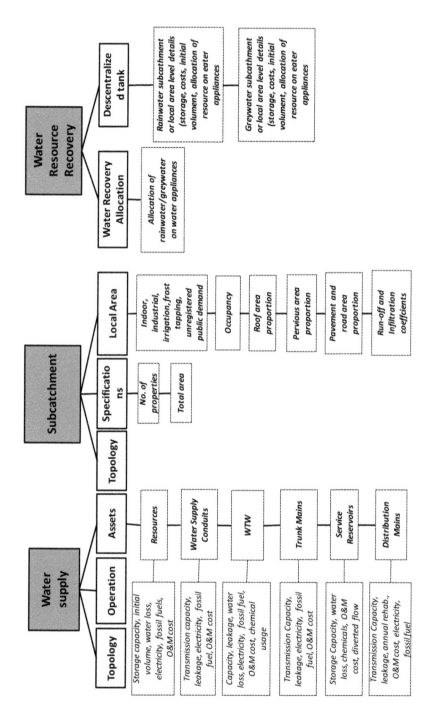

Figure 6.6- Main UWS components used in the WaterMet² model

Table 6.7-Input information for the WM2 model sub-catchment component

Component name	Unit	Puerto Ayora
Topology	Defined as only one sub-catchment area and one local area	
Number of properties	-	1996 (domestic)
Total area	Ha	163
Current indoor water demand	L/cap/day	163
Current Industrial/Commercial water demand	m³/day	1200
Average occupancy per property	Inhabitants/household	4
Roof area proportion	(%)	40
Pervious area proportion *	(%)	30
Pavement & road area proportion*	(%)	30
Run-off coefficient *	(0-1)	0.85
Infiltration coefficient *	(0-1)	0.9

*Values calculated based on literature review (d'Ozouville *et al.* 2008b)

6.8 WaterMet² model calibration

The WaterMet² model calibration in this study was done based on historical data of monthly water abstraction from the crevices serving as the water source of the UWS. The calibration parameters related to the capacity of the water resources in the WaterMet² model were adjusted by using the monthly records of water abstraction available at the municipal water department. The calibration was performed with historical groundwater abstraction rates (m³/day), which were divided in two periods: year 2012 for calibration and year 2013 for validation.

Table 6.8- Selected monthly coefficients for Puerto Ayora

Month	Puerto Ayora
January	1.027
February	1.019
March	0.991
April	0.993
May	0.967
June	0.965
July	0.922
August	0.941
September	1.014
October	1.050
November	1.036
December	1.080

Figure 6.7 shows a graphical comparison of the performance of the model by plotting simulated versus observed figures. Later, they were likewise validated with the following period (year). The statistical correlation coefficient (R^2) of 0.886 represents an acceptable value for this particular case study, regarding the lack of consistent data on the water pumping for supply, where daily records of water extractions for several days were missing. The value of the correlation is significant based on the acceptable ranges suggested by other similar works (Behzadian and Kapelan 2015). The model accuracy can be improved by increasing the amount of measured data used in calibration.

Based on the calibration and validation processes, the monthly variations for Puerto Ayora applied in WaterMet^2 (monthly coefficients of water demand profiles) are presented in Table 6.8. These coefficients were calculated based on the daily supply average for that particular month and the total daily average supply average for the years chosen for validation and calibration (2012+2013). These were applied for the entire planning horizon and used for both domestic and industrial water demand.

6.9 Alternatives and intervention strategies

Six potential alternatives were developed in this study. These alternatives vary from those aiming to increase water supply (e.g. desalination plant construction or RWH and/or GWR) to those aiming to reduce demand (e.g. leakage reduction, water meter installation or any other form of general water demand management). All of these alternatives have been identified with the aim to balance the long-term water demand. The detailed description of all alternatives is shown in Table 6.10. Leakage reduction and water meter installation are alternatives that have been proposed already over the last years, nevertheless, they have not yet been implemented (GADMSC 2012b).

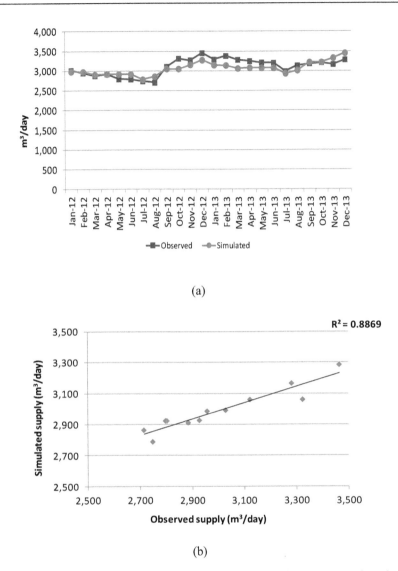

(a)

(b)

Figure 6.7- *Comparison between simulated and recorded supply as (a) time series of supply and (b) scatter plot for the years 2012 and 2013*

With this study we aim to quantify the impact of the strategies on the short and long-term. Furthermore, desalination with reverse osmosis has also been suggested due to popularity within authorities, since water quality issues (salinity) would also be improved. RWH has been proposed as a more sustainable option and due to the attractiveness in the smaller town of Bellavista. Moreover, GWR has also been proposed as a sustainable option as well and with the aim of reducing wastewater disposal, since submerged membrane bioreactor (MBR) offer a low-footprint with a high quality effluent for recycling domestic water (Verrecht *et al.* 2010). It is considered to be more feasible in-situ greywater recycling than the use of a centralized use of municipal wastewater, due to the quality of the wastewater. A treatment of municipal wastewater may involve more complex treatments and increase costs.

Finally, water demand reduction was also included, since the specific demand currently is considered high compared to other domestic demand in water scarce areas. Each alternative was analysed separately in order to assess the impact on the UWS. All of the previously proposed alternatives were discussed with the Department of Water and Sanitation of Santa Cruz. Some of them have already been proposed and some are new suggested alternatives.

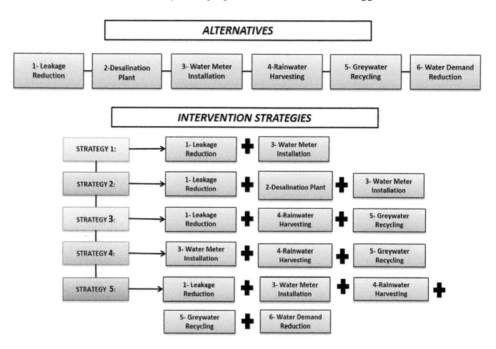

Figure 6.8-Intervention strategies applied to the UWS model

The proposed alternatives were first simulated individually in the WaterMet² model over the planning horizon (2014-2044) in order to analyse their impact to balance future water demand under different population/tourist growth scenarios. Table 6.9 shows the results of respective simulation model runs in terms of the fraction of the water demand delivered (i.e. covered) at the end of the planning horizon (year 2044), including the corresponding baseline at the start of the planning horizon (year 2014). Due to unsatisfactory (i.e. low) fractions of water demand covered by supply for most of the individual alternatives, these were combined to form five more complex intervention strategies (Figure 6.8). The complete description of all alternatives are shown in Table 6.10.

Table 6.9-Fraction of the water demand delivered

Population growth	Baseline	Alternative 1	Alternative 2	Alternative 3	Alternative 4	Alternative 5	Alternative 6
Slow	0.52	0.64	1.00	0.68	0.72	0.79	0.73
Moderate	0.35	0.36	1.00	0.37	0.40	0.43	0.41
Fast	0.17	0.17	1.00	0.18	0.21	0.22	0.20
Very Fast	0.10	0.11	1.00	0.11	0.16	0.13	0.12

Generally, the selected intervention strategies can start at any year of the planning horizon period. This study assumed that all the alternatives will be implemented starting at year 3, in order to give time to the municipality to implement and construct the different infrastructure needed for each of the proposed alternatives. The combination of strategies aimed to complement each other, and to improve the fraction of water demand delivered at the end of the planning horizon. Intervention Strategy # 5 is considered a combination of all sustainable alternatives, except the option with desalination.

6.10 Results and discussion

The current situation (baseline) and the selected intervention strategies were analysed and evaluated with respect to a number of key performance indicators (KPIs) for a 30 year planning horizon. The KPIs used here for comparison of the different selected strategies are total water demand, percentage of water demand coverage (i.e. fraction of water demand delivered), consumption per capita, energy consumption, and costs (capital and O&M), for each growth scenario.

Figure 6.9 shows the results of the Puerto Ayora case study. These figures portray results of year 30, as it has been considered as the most critical year, based on the population and tourism growth scenarios. Obviously, the most severe scenario is the very fast growth, which is driven by the governmental objective to optimise tourist revenues for the country.

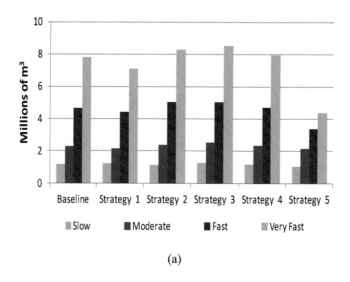

(a)

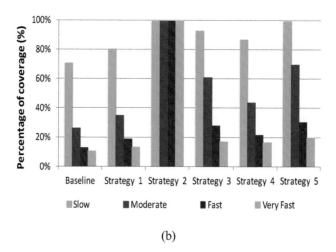

(b)

Figure 6.9- (a) Total water demand by 2045 for various growth scenarios and (b) Percentage of demand coverage by supply in 2045 for various intervention strategies in Puerto Ayora

Alternative	Description	Input values	Assumptions	Total Costs[b] (EUR/m³)	Reference
1) Leakage Reduction	Reduction from 28%[a] to 13% (1% annually).	Energy consumption: 0.66 KWh/m³ (current use of energy). The same values for all four growth scenarios	Installation of automatic and computerized leakage and control system (e.g. pressure and flow monitoring) and replacement of old pipes (17,800 m of PVC pipes).	0.66	Municipality of Santa Cruz and local providers
2) Desalination Plant	Installation of a new SWRO desalination plant (BWRO was not considered to avoid extra pressure on the basal aquifer and increase of salinity) with energy recovery system. Open seawater intake (35,000 ppm), 55% recovery rate, 99% salt rejection.	1) small growth (9,000 m³/day) 2) moderate growth (16,000 m³/day) 3) fast growth (28,000 m³ day) 4) very fast growth (50,000 m³/day) Energy consumption[c]: 3 KWh/ m³	Cost includes plant, land, civil works and amortization costs, chemicals for pre and post water treatment, energy requirement, brine dissolution and discharge, cooling towers (including electricity and steam), spares and maintenance (including membrane replacement every 5 years), and labour.	1) 1.27 2) 1.25 3) 1.23 4) 1.22	1) Ghaffour et al. (2013) 2) Al-Karaghouli and Kazmerski (2013) 3) Lattemann et al. (2010)
3) Water Meter Installation	Installation of water meters per premise with a rate of 10% annually.	140 EUR/unit (including installation and maintenance) The same unit cost for all growth scenarios	Installation of Flodis-single jet turbine device	0.04	Municipality of Santa Cruz

Table 6.10-Suggested alternatives for improvement of the UWS

4) Rainwater Harvesting	Installation of a household rainwater harvesting tank for indoor and/or outdoor use (2 m³)	Capacity calculated as 4000 m³ (approx. 2000 households) Energy consumption: 2 KWh/ m³	Water collected from roofs only[e]. The collected rainwater used for toilet flushing, hand and kitchen basin, showers and outdoor use. The cost includes purchase cost of tank, pumping, filters, flocculation, disinfection, delivery and installation, household plumbing, and mains water switching devices, energy consumption, maintenance and pump replacement (every ten years).	0.21	1) Tam et al. (2010) 2) Retamal et al. (2009) 3) Hauber-Davidson and Shortt (2011)
5) Greywater Recycling	Installation of single house on-site greywater treatment using a submerged membrane (MBR), including disinfection unit.	Based on household greywater treatment capacity of 350 L capacity and 2000 households; 5 inhabitants per household and 163 lpcpd. Flow capacity of 200 L/population equivalent Energy consumption: 3.11 KWh/ m³	Greywater collected from kitchen and hand basins and showers, which account to approximately 48% of total water demand. Household treatment assumed with membrane bioreactor system (biological treatment, aeration, and membrane filtration). Treated greywater used on-site for toilet flushing and outdoor use.	1.08	1) Fletcher et al. (2007) 2)Boehler et al. (2007) 3) Gnirss et al. (2008) 4) Fountoulakis et al. (2016)
6) Water Demand Reduction[f]	Reduction of specific demand of municipal water	Reduction from 163 lpcpd[d] to 120 lpcpd (assuming 1% annual reduction on water demand starting on year 3, in order to complete the reduction at the end of the planning horizon)	Assumed that the change of water tariff structure will reduce the average specific demand	-	-

[a] This value was considered for calibration purposes. [b] Total costs include investment costs, operations and managements costs, interest rate and extra costs and the municipality will assume all of them. [c] The cost of energy as observed in the literature ranges widely from 2 to 12 kwh/m³, however, since this would be a brand new plant we have selected a value towards the lower side. [d] lpcpd corresponds to liters per capita per day.

Based on Figure 6.9, it can be inferred that the current infrastructure would not suffice for any of the population growth scenarios. Even in the slow growth scenario, hardly 70% coverage of demand with supply would be reached, and in the very fast scenario hardly 20%.

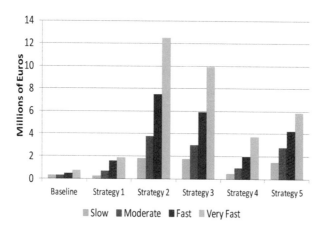

(a)

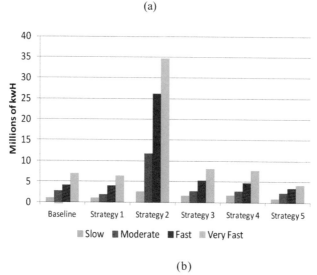

(b)

Figure 6.10- (a) Total costs on year 30 and (b) Total energy use on 2045 with different intervention strategies and growth scenarios in Puerto Ayora.

This also shows, that the current situation is not as perceived, since based on the volume of water supplied, the current coverage is calculated as 91%. However, the local community considers the coverage to be less (Guyot-Tephiane 2012). Figure 6.9(b) shows that only Strategy #2, which includes desalination, will fully cover the water demand by the end of the planning horizon, for all growth scenarios. This suggests that current growth trends are exorbitant and will generate a significant local population and tourism demand of water. Nevertheless, Strategy #5 (a combination of all alternatives, except desalination) will be sufficient, but only in the slow growth case scenario, which has been the one preferred by all NGO's and conservation authorities, but highly unlikely.

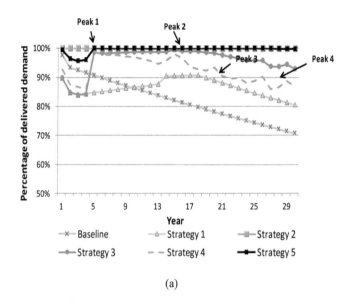

(a)

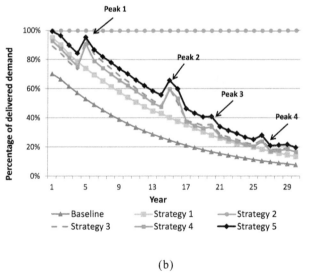

(b)

Figure 6.11-Coverage of demand with supply over the planning horizon for the (a) slow growth and (b) very fast growth scenarios in Puerto Ayora

Strategy # 2 increases water availability by installing a new desalination plant, complying with water demand over the entire planning horizon. Therefore, even though the other strategies have been suggested to avoid such an investment and greater potential environmental impacts on the island, they cannot meet the demand in 30 years. The best strategy to save on total water demand, reducing pressure on the supply system and infrastructure, is Strategy # 5, for all growth scenarios. This is because Strategy # 5 is the only one that considers and contributes to proper demand management, reducing current specific demand by 40 L/cap/day at the end of the planning horizon. However, is still insufficient for a 100% percent coverage of demand in year 2045 and therefore cannot cope with the future proposed growths; except on the slow growth scenario.

Regarding costs and energy use shown in Figure 6.10(a) and Figure 6.10(b), as expected, Strategy # 2 has significantly higher energy consumption and costs, than the other intervention strategies. This costs refer to the total unit costs for the total water demand produced at year 2045. The costs vary for each growth scenario, making the fast and very fast more unsuitable due to the enormous financial burden. This makes reference also to a higher investment (depending on the plant size calculated per growth scenario), as well as operation and management costs, implicated with a desalination plant and a much higher energy use for desalination treatment and process, compared to the other strategies, regarding this particular case study. Since this archipelago is located approximately 1000 km from the mainland, fuel for producing electricity needs to be imported from the mainland, adding extra costs to this option. Furthermore, GWR has also high costs due to pumping costs and investment per installed unit, nevertheless it is more environmentally friendly since it reduces wastewater and water demand. The most economical strategies are #1 and #4, which include leakage reductions, water meter installation, as well as RWH and GWR.

Even though Strategies #3 and # 4 are pretty similar because they both have RWH and GWR, the influence of water meter installation is more positive and therefore the costs reduce. On the other hand, Strategy # 5 has the higher costs after Strategy # 2, because of the combination of all the strategies, being a significant figure the alternative of leakage reduction (replacement of pipes and an automated leakage control system), as well as GWR. Regarding the energy consumption in year 30, Strategy # 2 is 3 to 4 times higher than the other strategies (4 KWh/m^3). Regarding energy use, the best option is Strategy # 5, because it decreases water demand. The other strategies do not seem to vary much on the energetic consumption compared to the current situation, since the total amount of RWH or GWR is not that significant when compared to the total water demand from the conventional municipal supply (the ones that will consume more energy).

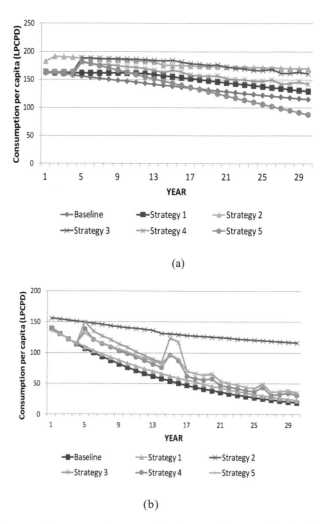

(a)

(b)

Figure 6.12-Consumption per capita over the planning horizon for the (a) slow growth and (b) very fast growth scenarios

Furthermore, Figure 6.11 shows the variations of water demand delivered to consumers over the planning horizon for different strategies and for the scenarios of slow and very fast growth only. As it can be observed from this figure, the first peaks on both diagrams start occurring in year 3, when the alternatives are implemented. The rest of the peaks can be explained by the influence of meteorological data, making some years better for RWH (and GWR) than the others. Therefore, based on historical precipitation rates, the methodology adopted predicts similar variations for the future, affecting directly rainwater collected by every individual household, making some years better than the others. Therefore, the amount of rainwater contributing to the water demand delivered is higher on the peaks. Strategy # 1's percentages of coverage have the tendency of decreasing over the years, more abruptly in the very fast scenario, suggesting that water meter installation and leakage reduction by themselves do not contribute significantly to solving the water coverage deficit.

On the other hand, Strategy # 2's percentages of coverage remain constant due to the increase in the transmission capacity and availability of water. In addition, Strategies # 3, # 4 and # 5 are influenced by rainfall levels, since rainwater is considered as an alternative in these three strategies. The variations between these last three strategies are due to the combination of other alternatives, regarding water meter installation, leakage reduction and water demand reduction.

Figure 6.12 shows the impact and evolution of different intervention strategies regarding the calculated per capita demand. This was calculated based on the prognoses of population growth for every year and for each scenario. In the slow growth scenario the specific demands for all strategies have more or less the same tendency to decrease at the end of the planning horizon, but not necessarily because of the reduction of consumption per capita, but due to the amount of total water divided by more people every year. The highest per capita consumption is observed for Strategies # 2 and # 3, which, as stated before, are the strategies that involve the increase of water availability. Moreover, Strategy #5 has a significant impact regarding the reduction of households' consumption and use, especially toward the end of planning horizon, where the per capita figure tends to decrease. In the case of very fast growth, the per capita consumption trends vary between strategies, reflecting each alternative selected and the type of population growth scenario. Unexpectedly, none of the strategies reduce per capita water consumption from the baseline scenario, but all of them will increase these figures over the years. Strategy # 2 seems to increase per capita consumption 2-3 times more when compared to other strategies, but this means that this strategy allows satisfying the customers completely without the need to reduce it. Nevertheless, these values reduce at the end of the planning horizon and stay within reasonable margins because of the high number of projected population for the latest years.

Also, current climate change impacts should be taken into account. This means that more extreme events of less or more precipitation may affect several strategies, such as the rainwater harvesting. Also, these possible events may affect even more the unbalance of the basal aquifer for the strategies that consider brackish water (strategies with leakage reduction, grey water recycling, water meter installation and per capita demand reduction). This may suggest that the quality of the water could reduce significantly, limiting its uses even more that in the current situation. Therefore, these strategies could be complemented by other individual alternatives, making them more complex. Furthermore, this suggests once again, that the desalination option needs to be considered if the future population growths targets the fast and very fast scenarios since this option would be inevitable.

6.11 Conclusions

This chapter addresses the issue of long-term water supply/demand balance for the main town of Puerto Ayora. To address this, a WaterMet^2 model capable of simulating the water balance over a 30 year period was built and calibrated. Five possible intervention strategies were defined by combining six individual alternatives, aiming at either increasing supply by using alternative water sources (e.g. desalination water plant construction, RWH and GWR) or reducing future water demand (leakage reduction, water meter installation and WDM). The impact of these strategies on the water system performance was evaluated by using suitable KPIs, under four scenarios of population growth. The KPIs appear to be very sensitive to the population growth, actual water demand (domestic and tourist) and leakage levels, which have been estimated based on other studies, but would need to be verified by further research.

Clearly, the current infrastructure would not be sufficient for any growth scenario, suggesting that fast and very fast growth scenarios are unsustainable and unaffordable. The results obtained show that the most viable strategy with respect to water demand coverage (i.e. fraction of future water demand met covered by supply) in the moderate, fast and very fast population growth scenarios is to install a desalination plant. However, this would increase the energy consumption drastically; exerting extra pressure on the current thermal plant and implying additional fuel importation from the mainland, increasing costs and negative environmental impacts. Moreover, the disposal of brine is likely to be a potential problem and this option requires high investment that most likely cannot be afforded by the local municipality. As stated by the study of Dhakal et al. (2014), desalination options are still the most energy intensive technology to produce drinking water and, most of the times is implemented as a last resort where conventional (freshwater) resources have been stretched to the limit. Furthermore, it produces considerable amounts of Green House Gas (GHG) emissions if fossil energy sources are used.

None of the suggested strategies would suffice for the moderate, fast and the very fast growth scenarios, expect for Strategy #2 (a combination of desalination, water meter installation and leakage reduction). Because of high potential environmental impacts associated with this strategy, a more sustainable option is to apply Strategy #5 (a combination of all alternatives, except desalination). However, if the annual population and tourist growths continue as the governmental objectives suggest, this strategy would not meet the expected future water demand. Furthermore, the uncontrolled water abstraction from crevices resulting from Strategy #5 may lead to intrusion of seawater into the aquifer compromising the water quality and making it unusable. As already identified by Pryet (2011), the current aquifer where La Camiseta crevice is located, has high infiltration potential, weak rainfall and probably negligible recharge. Finally, the WDM program suggested in Strategy # 5 would impact positively on the specific water consumption and hence would alleviate the need for additional supply.

A number of other analyses should be done in the future, which are out of the scope of this research. A combination of other alternatives could be further investigated, such as desalination for potable water (drinking and cooking) and the use of brackish water for all other requirements, in combination with RWH and GWR. Nevertheless, climate change could be a reason to reconsider RWH; if the supply system becomes highly dependent on this source, it would consequently be dependent on the weather. This also implies a further calibration of the meteorological data before decision involving rainwater can be taken with confidence. Finally, as observed in Strategy #5, the reduction of per capita water demand would have a considerable influence on the future supply/demand balance. Therefore, other WDM strategies (e.g. using water-efficient appliances and fitting especially inefficient toilets which account for 49% of total household demands should be further investigated in the future researches.

The WaterMet² methodology has been shown to be a useful and practical software for data limited and small case studies. The current study also demonstrated that WaterMet² can provide a holistic approach in modelling urban water systems under current and various future scenarios.

"Water and air, the two essential fluids on which all life depends, have become global garbage cans." — Jacques Yves Cousteau

7

MULTI-CRITERIA DECISION ANALYSIS OF WATER DEMAND MANAGEMENT OPTIONS FOR PUERTO AYORA

A Multi-Criteria Decision Analysis (MCDA) is tested in this chapter as a suitable methodology which leads to a set of alternatives that aim to mitigate the future water supply shortage. This was carried out to reveal the most optimal solution in terms of environmental, technical, economic and social criteria by using the DEFINITE software. Different preferences were assessed from four groups of stakeholders based on their evaluations regarding each criteria and indicators. Also, an uncertainty and sensitivity analysis were conducted, which evaluated how the ranked alternatives may change, if the weights and scores for each indicator were to be altered. The results indicate that the 'best' alternative for the majority of the stakeholders' sessions is either desalination or the option combining greywater recycling, specific demand reduction and rainwater harvesting. For the decision makers and experts group, the installation of a desalination plant is the preferred option, ensuring 100% coverage of the total water demand at the end of the planning horizon (year 2045), but with significant environmental threats and high costs.

This chapter is based on:

Reyes, M., Petricic, A., Trifunovic, N., Sharma, S., and Kennedy, M. (2017) A Multi-criteria Decision Analysis of Water Demand Management Options for Puerto Ayora-Santa Cruz Island (Galápagos). *(Paper submitted to Water Resources Management Journal)*

7.1 Introduction

Many tourist islands have scarce information regarding their water demand and supply. Therefore, for many of these islands, water resources planning and management is particularly lacking due to the absence of proper data. In order to ensure the coverage of future water demand with supply, different intervention strategies need to be proposed, and further evaluated from several relevant perspectives.

The MCDA is an often-used tool to carry out analysis for decision making purposes. It is an integrated assessment, which is a form of combined sustainability evaluation (Wang *et al.* 2009). By this methodology, the intervention strategies are evaluated taken into consideration involved (relevant) stakeholders' preferences and points of view. This will contribute significantly to strategic planning and water resources management in islands, and the study presented in this chapter for Santa Cruz Island could serve as a template for other tropical islands undergoing similar water issues.

The methodology carried out in this chapter aids to propose solutions for future challenges regarding optimal water balance, overcoming water scarcity caused by high tourism rates, and also to preserve such places for future visitors. Because Puerto Ayora is the major touristic centre on the Galápagos, applying an MCDA here serves as a template for other touristic islands, providing a basis for the development of criteria, as well as specific indicators.

This chapter elaborates on the assessment of the five previously-developed intervention strategies proposed to solve the future water deficit in Puerto Ayora (Chapter 6). The potential strategies were assessed from the environmental, social, technical and economic perspectives, by using the DEFINITE software developed by Janssen *et al.* (2001). The final results include the 'best' alternative considering all the criteria, based on the selected stakeholders' preferences. The sensitivity and uncertainty analyses were carried out as well, with the aim of analysing the effects on the final results, based on the variation of initially adopted scores and weights.

7.2 Multi-criteria decision analysis

The MCDA encompasses an integrated and complete assessment of multiple suggested alternatives, with a set of tools for any decision making process. These analyses comprise of applied mathematical algorithms that evaluate a collection of different values and factors. The evaluation is usually carried out on problems with conflicting goals, high uncertainty, different forms of data and information, multiple interests and perspectives, and complex biophysical and socio-economic systems (Wang *et al.* 2009). The ultimate goal of this approach is to define the most feasible and sustainable solution of a certain issue at a low cost and considering all preferences of the participants (Linkov *et al.* 2006).

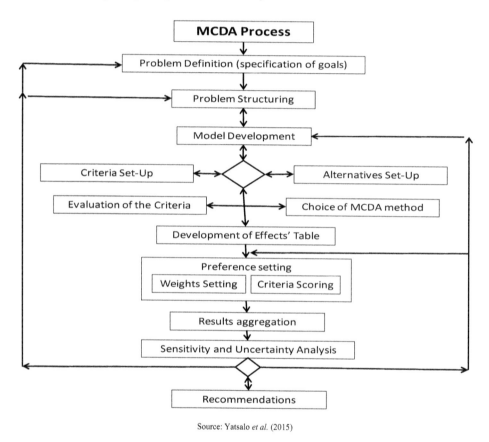

Source: Yatsalo *et al.* (2015)

Figure 7.1- Schematic of the procedure of the MCDA

Even though these types of analyses have been considered as subjective, the use of MCDA method has increased rapidly since the 1990s, especially with environment related studies and researches. This has been due to the increased complexity of projects, higher level of public and stakeholder participation, and the need of management of information in a more transparent way (Salaguste-Anarna 2009). The use of MCDA provide a reliable method, which allows to rank different proposed alternatives in the presence of numerous objectives and constraints (Polatidis *et al.* 2006). Despite the large number of available MCDA methods, none of them is considered the best for the different decision making situations (Simpson 1996, Guitouni and Martel 1998, Salminen *et al.* 1998). Instead, it is necessary to identify techniques and approaches that fit better to a certain situation or not. Thus, MCDA methods have been tested to be ideal for water resources management and planning, and are usually preferred to other methods for reasons that include transparency and accountability to decision procedures (Mutikanga *et al.* 2011).

Moreover, one of the advantages of the MCDA is the capacity to involve several stakeholders. This allows the integration of the decisions from different groups, involving stakeholders with different perspectives, resulting in a more thorough understanding of the points of view held by the involved parties (Kiker *et al.* 2005). Also, it allows decision makers to view solutions clearly to a previously defined problem, and also helps policy makers to involve different criteria within new policies.

Several steps were followed in order to carry out a MCDA of the proposed alternatives for ensuring the future water demand coverage in Puerto Ayora. Figure 7.1 shows the generic process, based on Yatsalo *et al.* (2015). The most important step concerning the MCDA is the proper formulation of the problem, as well as a clear definition of the extent and level of involvement of the participants and proper evaluation (Franco and Montibeller 2010).

7.3 Research methodology

The steps followed for the MCDA are described below.

7.3.1 Criteria definition and alternatives' selection

The problem was defined as the water supply deficit in the town of Puerto Ayora in the year 2045. The key objective of the analysis was to find a suitable and sustainable solution. The five intervention strategies shown in Table 6.10 and Figure 6.8 were obtained from Chapter 6. Therefore, the results for the end of the forecast period (2044) were the basis for the input data for the MCDA. Four main criteria were selected for this study: (1) environmental, (2) technical, (3) economic and (4) social. Each of these criteria was further described with relevant and measurable indicators, which allowed evaluating the impact and performance of each strategy under the defined criteria. The DEFINITE software was selected as the tool for this analysis, since it can be used on a wide variety of problems, regarding different disciplines. This software was developed to help improve the quality of decision, by methodical procedures which lead experts through a number of interactive assessment sessions. It uses an optimization approach to integrate all information provided by the stakeholders involved to a full set of value functions leading to a scientific based alternative (Janssen and van Herwijnen 2011).

7.3.2 Effects table, scoring and standardization

After the definition of criteria and its suitable indicators, each strategy was scored under each indicator. This scoring, which is denominated as the 'effects table', refers to the assessment of the performance of each intervention strategy against all the pre-defined indicators in a qualitative or quantitative way. This step is the core of the MCDA, representing the input information for the DEFINITE. Furthermore, every strategy needs to be evaluated as accurately as possible, for every indicator and criteria.

The data used for scoring and populating the effects table were obtained from Chapter 6, including the KPIs of water demand, water losses, energy use and costs. In the current chapter, only the results from the moderate growth scenario were used (4% annual growth for tourists and 3% for local population). The missing information was taken from the literature, such as potential waste quantities from the different selected strategies, as well as laws and regulations used locally. Other information was also taken from interviews with local experts, such as for the social criteria and the technical one.

Table 7.1- Score symbols for qualitative assessment

Symbol	Meaning
---	very large negative effect
--	large negative effect
-	small negative effect
0	no effect
+	small positive effect
++	large positive effect
+++	very large positive effect

Source: Janssen and Herwijnen (2011)

The scoring for indicators was done using different types of scales/units. First, the indicator values were defined as qualitative or quantitative, in order to assign the scale/unit that will be used. The scales/units used were: (1) ratio, (2) interval, (3) ordinal, (4) binary scale, and the (5) ---/+++ scale. The ratio scale refers to the proportionality of values, the interval scale portrays the ranges of amounts, the ordinal scale ranks the effects of an strategy against certain indicator, the binary scale indicates whether the effect does or does not occur and the ---/+++ scale estimates the qualitative values. Table 7.1 explains the meaning of the last mentioned scale.

Table 7.2- Criteria, indicators, units/scale, standardization method and ranges selected for the MCDA for a better water supply system in Puerto Ayora

Indicator	Cost/Benefit Correlation	Unit/scale	Standardization Method	Minimum Range	Maximum Range
ENVIRONMENTAL CRITERIA					
Land use	C	m²	Goal	0	10,000
Discharge of wastewater	C	m³/day	Goal	0	50,000
Seawater intrusion	C	Ordinal	Exponential value	2	5
Energy consumption	C	kWh/ m³	Maximum	0	3
Chemical use	C	Binary	Yes=0, No=1	No	Yes
Impact on endemic species	C	Ordinal	Exponential value	1	5
Impact on marine/land ecosystems	C	Ordinal	Exponential value	1	5
TECHNICAL CRITERIA					
Improvement on hours of service	B	Binary	Yes=0, No=1	No	Yes
Coverage of demand with supply	B	% of demand	Goal	30	100
Water losses	C	% from water produced	Goal	9	28
Robustness of the WS system	B	Ordinal	Exponential value	1	5
O&M of the WS system	B	Ordinal	Exponential value	2	5
Alternative water sources contribution to overall balance	B	% annually	Goal	5	50
Compatibility with the existing system	B	0/++	Maximum	0	++
ECONOMIC CRITERIA					
Capital cost	C	M €	Maximum	0	21.6
O&M cost	C	M €/year	Maximum	0	5.4
NRW income generation	B	€/year	Maximum	0	312,412.16
WDM income generation	B	0/++	Maximum	0	++
Employment generation	B	0/++	Maximum	0	++
Increase in water tariffs	C	--/0	Maximum	--	0
Increase in tourist capacity	B	# of tourists	Goal	7,335	15,000
SOCIAL CRITERIA					
Social acceptability	B	0/+++	Maximum	0	+++
Willingness to pay	B	0/++	Maximum	0	++
Transparency on project implementation process	B	0/++	Maximum	0	++
Water quality improvement	B	0/++	Convex	0	++
Annual infection and other water-related diseases risk	C	--/0	Convex	--	0
Compatibility with current legislations	B	Binary	Yes=0, No=1	No	yes

After defining the scale/unit with which each indicator will be assessed, it was necessary to determine the nature of each indicator as Cost (C) or Benefit (B) relation. Cost refers to the indicators which have negative correlation between the score and the effect, suggesting that the higher the score of the indicator, the worse the effect produced. On the other hand, benefits (B) refer to positive correlation, the higher the score, the better the effect produced for the analysis. Later, the effects table was populated with the scores assigned to each indicator, corresponding to every intervention strategy.

Afterwards, the scores assigned to each indicator were standardized, since the values attributed in the effects table were not yet comparable and the units were not uniform. Therefore, each indicator was standardized with a unit-less value between 0 and 1. For this, different options available within the software were used to convert the original indicator scores (Janssen and Herwijnen, 2011). The methods used for this MCDA were: the maximum method, the goal standardization, the convex function and the yes/no standardization.

Each method was analysed and carefully selected for each indicator. Therefore, the most suitable standardization method, as well as value function, was used, selecting the one that will adjust better to the scale of the effects. Hence, the chosen method provided the desired and targeted value for each indicator representing the water supply system at the end of the project horizon. Table 7.2 presents the selected criteria, indicators, cost/benefit relation, units/scales, the standardization method used, as well as the ranges of scores for each indicator.

7.3.3 Weight Allocation

Following the standardization step, the weight allocation for every criteria and indicator was done. The weights were obtained from different stakeholders' preferences. For this, a questionnaire was distributed to 37 previously-selected stakeholders, clustered into four different categories as shown in Table 7.3.

The questionnaire had six questions regarding valuation and importance of the criteria and its indicators, rating them from 1 (the least important) to 5 (the most important). Later, the answers from each stakeholder group were processed, in order to arrive to the weight allocation for each criteria and indicator (Figure 7.2.)

The results of the weights valuation were then calculated based on the average of the respondents belonging to each group of stakeholders, carrying out a different MCDA session for each stakeholder group. This was the key part for the ranking of alternatives in the MCDA.

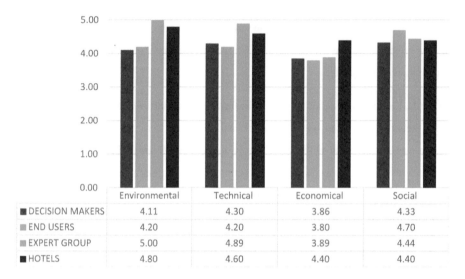

	Environmental	Technical	Economical	Social
■ DECISION MAKERS	4.11	4.30	3.86	4.33
■ END USERS	4.20	4.20	3.80	4.70
■ EXPERT GROUP	5.00	4.89	3.89	4.44
■ HOTELS	4.80	4.60	4.40	4.40

Figure 7.2- Ranked preferences of selected stakeholders based on distributed questionnaire in Puerto Ayora

7.4 Results of the MCDA sessions and discussion

7.4.1 Distribution of weights based on stakeholders' input

The aim of this step was to standardize the results from stakeholders' responses into values from 0 to 1, using appropriate methods included in the DEFINITE software. For the criteria, defined as weight level 1, the direct weighting method was chosen, because it assigns quantitative weights directly, according to the numerical input used by the stakeholder feedback. In the direct weighting, the sum of the weights of the criteria for each particular stakeholder session, must be equal to one. Moreover, the indicators were defined as weight level 2, where the expected value method was used. This method ranks the effects in order of their importance, assigning quantitative values to each effect (score assigned). Some effects may have the same ranking, which means they have equal importance in the analysis.

Institution	Stakeholder position
Expert's Opinion	
Charles Darwin Foundation	Director
International Conservancy	Manager
World Wildlife Foundation (WWF)	Director
Island Conservation	Director
Galápagos Conservancy	Coordinator of projects
FUNDAR (Alternative Sustainable Development Foundation)	Executive president
San Francisco University of Quito (USFQ)	Professors (2) in Environmental Engineering
Galápagos Science Center (GSC)	Director
Polytechnic University of Ecuador (ESPOL)	PhD Fellow on groundwater in the Galápagos Islands
USFQ-GSC	Professor on Human Ecology-Co-director of GSC
Decision Makers	
Galápagos Governmental Council (CGG)	Planning Coordinator & Ministry's Advisor in Environmental Subjects
Direction of the Galápagos National Park (DPNG)	Director of Environmental Management & Researchers (2) on Water Quality
Municipality of Santa Cruz	Mayor's Advisor, Planning Director, Chief of Water and Sanitation Department & Coordinator of Environmental Management
Ministry of Tourism (MINTUR)	Tourism Observatory Program Manager
Investment Fund for Introduced species	Director
Municipality of Santa Cruz	Galápagos Councillor
Domestic End Users	
10 random selected families	
Tourist End Users	
5 random selected hotels	

*Stakeholder's have different sizes due to the number of received responses.

Table 7.3-Stakeholders selected for weigh allocation of criteria and indicators in Puerto Ayora.

Table 7.4-Weight allocation for main criteria for Puerto Ayora

Stakeholders	Environmental	Technical	Economic	Social
Experts	0.273	0.273	0.212	0.242
Decision Makers	0.251	0.240	0.240	0.270
Domestic End Users	0.250	0.250	0.220	0.280
Hotels	0.263	0.253	0.242	0.242

Then, the total weight was calculated based on the product of weight level 1 and weight level 2. This result became then the actual weight for the MCDA evaluation. Table 7.4 illustrates the standardized weight allocation for the main criteria, according to the stakeholders' preferences.

7.4.2 Ranking of the alternatives

The software has several methods which can be used for the ranking step of the MCDA: the Weighted Summation, Electre 2, Evamix and Regime. For this step, the weighted summation method was chosen, which is based on the MAUT (Multi Attribute Utility Theory) model. This method was selected because it is considered to be the most appropriate, since it is reliable, straight forward and transparent (Janssen *et al.* 2001, Herwijnen and Janssen 2004). This method uses the effect scores to rank the proposed intervention strategies, processing the standardized table into a ranking of alternatives. A total of four sessions were conducted and all alternatives were ranked using the same method in order to facilitate comparisons and conclusions. The results of the ranking of the alternatives are illustrated in Figure 7.3.

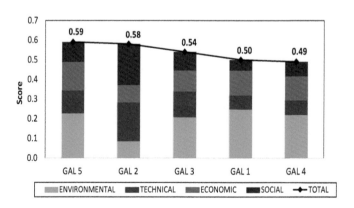

(a)

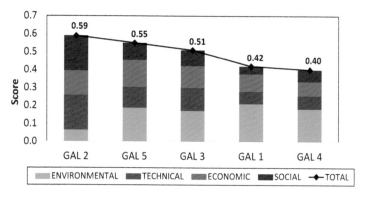

(b)

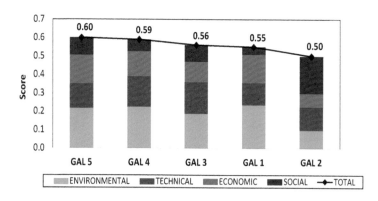

(c)

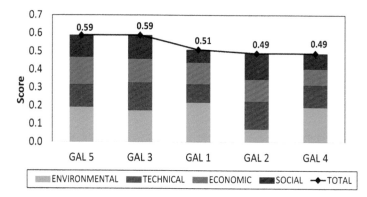

(d)

*Figure 7.3- Ranking of alternatives based on (a) Experts, (b) Decision makers, (c)
Domestic end-users and (d) Hotels preferences*

Figure 7.3 shows that Gal 2 alternative prevails for the decision-maker's group, due to the largest contribution from the technical and social criteria (the only alternative covering the total water demand at the end of the planning horizon, and also the only one that significantly improves water quality). For the experts group, Gal 5 alternative is the preferred one, due to the largest contribution from the environmental criteria. However, Gal 2 follows closely the option ranked first. For the domestic end-users session, priority was given to Gal 5, mainly due to the highest contribution from the environmental criteria, followed by the economic criteria. Surprisingly, for the hotel end-users, a significantly high preference is given to the environment, grading Ga l5 as the first option, and not the desalination option, as expected. Gal 5 scores reasonably well in all of the defined criteria, resulting in the most consistent distribution of ranking across all of the sessions, taking the first or the second position. On the other hand, for the domestic end-users, Gal 2 is always one of the last options.

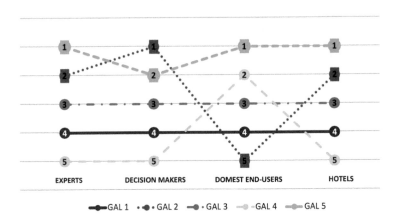

Figure 7.4-Comparison in rankings of MCDA results for Puerto Ayora

Furthermore, Gal 3 tends to be the third alternative in the overall ranking in all of the sessions. The highest influence for Gal 3 belongs to the environmental and technical criteria. Because it encompasses lower water supply system improvement measures than Gal 5, its performance is recognized by being somewhat poorer, resulting mostly on the third place. On the other hand, alternatives Gal 4 tends to be the last option for most sessions, expect for domestic end-user, where is positioned as second.

In general, for every analysis where either technical or social criteria have more influence, Gal 2 alternative is ranked higher, while Gal 5 takes the lead when environmental criteria have higher weighting.

The ranking summary is shown in Figure 7.4. As observed, the alternative that is ranked the highest and shows more stability is Gal 5. Occupying first and second places, followed by Gal 3, which is ranked always third. On the other hand, Gal 2 is first for decision makers group of stakeholders, but also takes the last place for the local population session, capturing the noticeable different preferences, especially regarding the environmental criteria and indicators. These results show that for local population as well as experts (which involved NGO's and researchers), the preservation of the environment is more important than the stakeholders involved with decision making.

7.4.3 Uncertainty analysis

The DEFINITE software can also assess the sensitivity of the ranking of alternatives, by varying the effect scores and assigned weights of the indicators. In order to evaluate the influence of uncertainties to a lower or higher extent of the sensibility and the reliability of the results, the percentages of the effect scores were examined with a ±50% variation. This was done based on the fact that some of the input data used to populate the effects table was assumed.

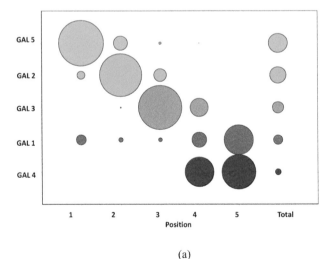

(a)

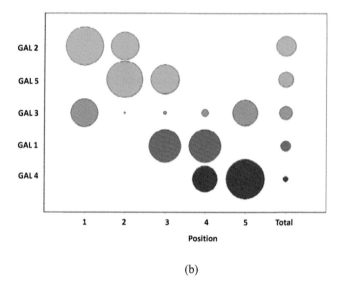

(b)

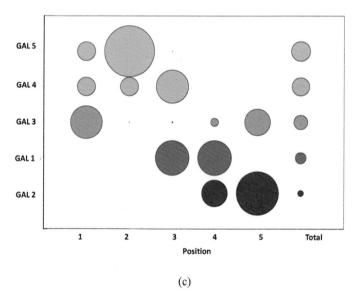

(c)

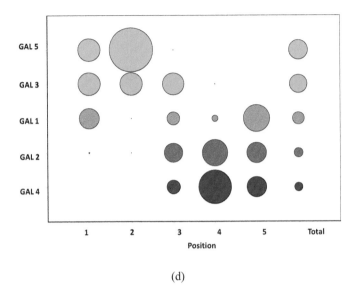

(d)

Figure 7.5- Probability of alternative ranking with 50% uncertainty score for (a) Experts, (b) Decision makers, (c) Domestic end-users and (d) Hotels sessions.

Since some results showed small difference between the rankings (0.1 to 0.2), the uncertainty analysis was carried out in order to examine the impact on the ranking, when the effect scores were changed. This also reflected the consequences of higher- or lower population growth impacts, since this analysis was done using only the moderate growth scenario. With this uncertainty analysis the other population growths used in the previous chapter could be also tackled. The weight uncertainties were not taken into consideration in the uncertainty analysis, since these have been determined from the distributed questionnaires, which are assumed to be 100% accurate.

Figure 7.5 shows the probability of an intervention strategy to change its ranking, if the scores are varied. For instance, Figure 7.5(a), which shows the experts group, explains that Gal 5 alternative has a high probability (72%) to keep the first place, as well as Gal 2 and Gal 3 to keep their original ranking. On the other hand, Gal 5 has low probability to take the second or any other position. The same consistency happens for the rest of the alternative sin this session has a considerable probability to take the fourth place. This suggests that because this group included NGO's, environmentalists and researchers who are aware of the importance of the ecosystem, Gal 5 will remain on the first position. The values of the effect scores are therefore not very sensitive, reflecting that there would not be an alteration in the overall ranking by a small change on the effects scores.

Regarding the decision makers session (Figure 7.5b), the results show relatively stable uncertainties, and is very similar to the previous session. The slighter smaller size of the circles compared to the precious sessions, suggests the probability is not that high for the ranking to change. This means that Gal 2 alternative will always be ranked first or second, as well as Gal 5 (as in the original ranking), with low or even null probabilities to take the rest of the other positions. This suggests that the given effect values would have to change considerably to a higher or lower extent, in order for Gal 5 (originally ranked second) to take the first place. There is not a significant discrepancy for this session in any of the options, since no option ranked last has probability to end up first or second or vice-versa. Therefore, the desalination option is the favourite one for this stakeholder group, since they give preference to the water supply system reliability and public health indicators (demand coverage, and water quality improvement), over environmental protection or cost criteria. The possibility of increasing tourist capacity, as well as 100% of the demand coverage at the end of the planning horizon, are advantages of the Gal 2 option, portraying that the desalination option would still prevail even if the scores are changed drastically. This suggests that this group is not very inclined to more sustainable measures.

Furthermore, the domestic end-users session shows a more unstable ranking, since Gal 5 which was originally ranked first, has high probabilities to end up second. Even though this group has the same values of preferences for the environmental and technical criteria, the desalination option has high probabilities to keep the last place and slight probability to be in the fourth position. This stakeholder group has a strong preference for options without extra environmental pressure, and the cheapest options tend to be on the second positions. Furthermore, this group gives extra emphasis to environmental preservation and WDM measures, even with ±50% of score uncertainties. Furthermore, the water quality and demand coverage seems to be of lesser concern for this group.

The hotels session in Figure 7.5(d), also shows also an unstable ranking, showing that Gal 5 has a high probability to take the second place, while Gal 3 and Gal 1 has stable probabilities to keep the original positions, but not for the last two places in the ranking. Regarding Gal 2 and Gal 4, they would tend to compete for the last places. Due to the environmental criteria and social contribution, the options that could be ranked first are the most sustainable ones, therefore

Gal 2 will keep the fourth position or end up last. Surprisingly, this stakeholder group has almost the same weight allocation as the previous one, despite the fact that this group is a major consumer. Furthermore, this group is still concerned about the environment and the negative impact of possible installation of a desalination plant.

In conclusion, the most robust alternatives are Gal 2 and Gal 5 keeping their original positions (first or second) in most of the uncertainty analysis sessions. Furthermore, Gal 3 and Gal 4 were moderately sensitive to the defined uncertainty variations because they often changed the ranking, usually for one position higher or lower depending on the stakeholder preferences. Therefore, all of the options would never tend to have a dramatic change in their original position. Finally, Gal 1 was the alternative with the highest level of ranking uncertainty. In most of the analysed sessions, this alternative competed for almost every position, originating from the wide uncertainty assumption.

Some of the preferred alternatives do not completely cover the water demand at the end of time horizon of 30 years, but they assume lower environmental impact, lower costs, and lower water tariffs (except desalination). If any of these alternatives would be adopted, this would mean that a large tourist expansion in Puerto Ayora is not possible, as most decision makers prefer (experts, domestic and hotel end-users). Thus, stakeholder group in charge of creating new policies and making decisions, do not prefer a positive long-term preservation of the ecosystem, since Gal 2 would never end up last.

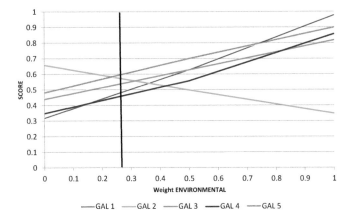

(a)

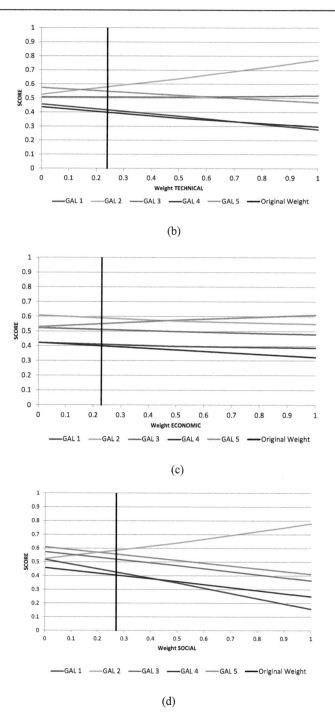

(b)

(c)

(d)

Figure 7.6- Sensitivity analysis of weight allocation on (a) environmental,(b)technical, (c)economic and (d) social criteria for the Experts session

7.4.4 Sensitivity Analysis

The final analysis encompassed the sensitivity analysis of the main criteria group weights provided by the selected stakeholders. Figure 7.6 and Figure 7.7 show the results only for the experts and decision makers sessions, respectively. Only these two groups were considered for this analysis, since we assumed they have better knowledge on the current situation and possible impacts of proposed measures.

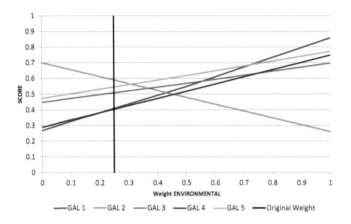

(a)

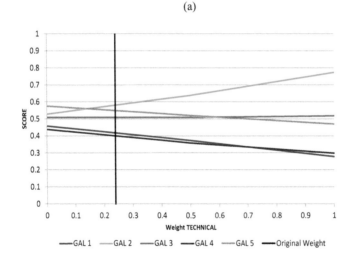

(b)

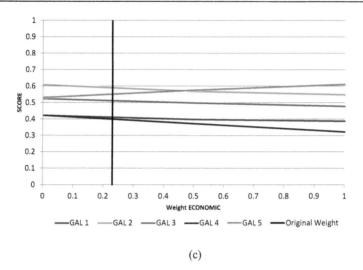

(c)

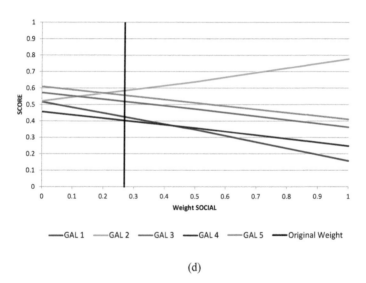

(d)

Figure 7.7- Sensitivity analysis of weight allocation on (a) environmental, (b)technical, (c)economic and (d) social criteria for the Decision Makers session

In Figure 7.6 and Figure 7.7 steeper slopes are observed in some of the graphs than in others. A steeper slope means that the criteria are more sensitive to a minor change on the weight, influencing greater on the final ranking of the alternatives. The X-axis indicates the extent of variation of the weight, and the vertical line is the original weight provided by the stakeholder's feedback. On the other hand, the Y-axis indicates the original score obtained by each alternative in the original analysis.

Based on these results, the experts group shows less sensitivity in the ranked results than on the decision makers group. For instance, the first ranked alternative (Gal 5), when analysed under the environmental criteria, remains on the same place within a wide range of weight variation. It would rather lose it if the weight lowers (original 0.26) and would be therefore replace by Gal 2. Moreover, if the weight value would change to 0.7 or less, the first place would be taken by Gal 1 (including leakage reduction and water meter installation). This suggests that this criteria is not very sensible to a small change on the weights, and if it becomes higher, the preference is still given to options without desalination. However, under the technical criteria weights, the situation is different, since the first position originally belonging to Gal 2, will stay on the first place if the original weigh (0.23) is increased up to 1, suggesting that if more importance is given to the quantity demanded and satisfied with supply, as well as water quality, no other alternative will replace the option including desalination. Nevertheless, if it decreases by only 0.03, it would be replaced immediately by Gal 5. Regarding the economic criteria, the lines are less steep, which means they are less sensible to the change in the weight provided by any stakeholder. This means that the results would be practically the same if this weight is altered significantly. However, for the other criteria weights, small changes can alter the whole ranking, giving it always advantage to Gal 5, which includes all solutions proposed, except desalination. For this group, even though customer satisfaction is considered (which would be provided by Gal 2), a small change on their feedback means that a more sustainable option would be preferred.

Regarding the decision makers group, the Gal 2 alternative loses its advantage over Gal 5, which was ranked first. Under the environmental criteria sensitivity, Gal 2 alternative is ranked in the first place only until the weight value increases by 0.5. Within the range from 0.32 to 0.65, Gal 5 alternative would be ranked first. With the values of environmental criteria below 0.26 there is a steep inclination of the desalination alternative, meaning that it is firmly positioned on the first place. Regarding the technical criteria, it is more sensible and could be, by a very small change, replaced by Gal 5. Therefore, over this weight, the desalination alternative (Gal 2) provides the best results in technical criteria group, especially regarding the indicators of coverage of demand with supply (100%), improvement of hours of service (continuous water supply), and robustness of the water supply system. Since it is the only alternative that significantly increases water supply, the higher the weight allocation of technical criteria, the more it stabilizes this alternative on the first place. On the other hand, lower values of technical criteria, switches the rank in favour of alternatives with higher scores in environmental criteria. For the economic criteria, the sensitivity is low, which is shown by much less steep lines. Since Gal 2 is the most expensive alternative, desalination is the preferred option under the economic criteria only up to 0.3 values (the original value is 0.23). Finally, regarding the social criteria, when the weight value drops below 0.15, Gal 2 loses its advantage over alternative Gal 5. Therefore, the social criteria can be considered an important one, since once more; a small change can alter all the results. This criterion includes paramount public health indicators, and those can be improved only by the desalination option. Nevertheless, this alternative can be easily substituted with less environmentally hazardous and cheaper options with small alteration of the weight scores of the main stakeholders' groups.

7.5 Summary and conclusions

The MCDA methodology has proven to be a suitable decision support tool, which may provide a thorough analysis regarding future water supply and demand options for tourist islands, under various growth scenarios. Also, it provides clear results under pre-defined indicators, which will aid decision makers and relevant authorities to make a scientific and supported decision to confront the future water crisis. However, the indicators, as well as their original values and ranges, have been proven to be case dependent and case sensitive.

In this chapter, different sets of measures for improving the water supply system of Puerto Ayora were analyzed with four main groups of criteria. The aim was to obtain the most sustainable solution for mitigating the future population and tourist growth. Also, it aimed to analyze the alternative which will provide an optimal balance between water supply and demand for the future conditions, with lowest impact on the fragile ecosystem, and with the most affordable cost.

As results showed, the Gal 2 and Gal 5 alternatives were ranked on the first position by the different stakeholder groups. Gal 2, which includes desalination, was preferred by decision makers. On the other hand, Gal 5 was preferred by experts and the two groups of local population (domestic end-users and hotels), which included all of the options, except desalination. These differences in the results can be attributed to the technical and environmental preferences given. Where the technical criterion has more weight, the desalination option tends to take the lead. However, if more weight is given to environmental criteria, Gal 5 takes the first position. Furthermore, based on the sensitivity analysis, Gal 2 tends to lose the first position easily by small changes on the weight values, because Gal 5 portrays a moderate environmental impact (low levels of wastewater discharge, lower impact to environment and sea water intrusion), moderate costs of implementation and operation and maintenance, but only 60% of water demand coverage at the end of the project horizon. On the other hand, Gal 2 is the only alternative that guarantees 100% coverage of water demand with supply, as well as improvement of the water quality to meeting drinking water requirements at the end of the project horizon (in 30 years). Therefore, decision makers give preference to this option. Nevertheless, despite these obvious advantages, due to the higher costs and negative environmental impacts identified, it can easily be replaced by Gal 5 in most of the sessions, suggesting the consideration by the different participants of this type of fragile ecosystem.

Regarding the uncertainty analysis, Gal 2 and Gal 5 have the highest probability to be ranked first or second in all of the sessions, as well as for the last ranked options, have probability to take the fourth or fifth place, but never the first place. This means that the results are certain and would not change drastically even if the effects scores are increased or decreased by 50%. As for the sensitivity analysis, it shows some criteria to be more sensitive than others. For instance, if more priority is given to environmental criteria, Gal 2 could never take the lead because of the negative impacts on the environment. This suggests that stakeholders prefer less quality, continue with the current situation, with little improvement in more sustainable terms, than installing a desalination plant. Moreover, it also showed that both groups analysed (decision makers and experts) have high sensitivity to small weight changes, since Gal 2 and Gal 5 may easily be replaced by other alternatives. Moreover, the technical, social and environmental criteria showed higher sensitivity than the economic criteria, based on the steepness of the lines, suggesting that if the priorities of costs are changed, it would not have an impact on the final results, and that priority is given by stakeholders to other factors.

Even though this analysis was done with just a moderate growth scenario, the total coverage of the demand according to the results from the WaterMet^2 software at the end of the project horizon would be limited to 60%. Consequently, other alternatives should be considered in conjunction with the desalination option, especially concerning the water quality improvement. For instance, a dual water supply could be explored, where the drinking fraction of the demand could be covered by desalination and the rest by the current brackish-water system. Moreover, the consumption at household level could be reduced by introducing a rational water tariff structure. Finally, the suggested tourist growth should be limited, and the present trend of tourist arrivals should not be continued. It would be necessary to adopt a minimum threshold value of the demand coverage for each of the assessed alternatives, and according to that then develop alternatives. Also, it would be interesting to assess the full potential of the RWH, as a centralized system with more detailed studies.

Finally, more studies would be needed to arrive at more accurate values of certain quantitative indicators. Those studies would need to encompass various types of long term modelling (hydraulic, hydro-geological, physical, demographic and economic impact, etc.). More detailed determination of social acceptance criteria is needed as well, in order to come up with proper descriptive values for the qualitative indicators. Appropriate methods would encompass public surveys, workshops, meetings at community levels, lectures with feedback, public discussions etc.

"Nothing is softer or more flexible than water, yet nothing can resist it."

- Lao Tzu

8

HYDRAULIC MODELLING OF DEMAND GROWTH IN PUERTO AYORA WATER DISTRIBUTION NETWORK

This chapter elaborates the hydraulic characteristics of the water supply network of the town of Puerto Ayora. First, it intends to replicate the household individual storage by simulating nodal tanks with the use of the EPANET software. Later, it uses the Pressure-Driven Approach (PDA) to develop a methodology that estimates the overflow of storage facilities, one of the main sources of wastage in Puerto Ayora. Finally, it uses the Demand-Driven Approach (DDA), with the aim of assessing the network in the future, under four population growth scenarios. With the chosen moderate growth scenario, two options are suggested in order to tackle the water supply issues at the end of the planning horizon.

This chapter is based on:

Reyes, M., Trifunović, N., Sharma, S., and Kennedy, M. (2017). Hydraulic modelling of demand growth in tourist islands, case study: Galápagos, Ecuador. *Paper submitted to the 15ᵗʰ international Water Management and Hydraulic Engineering.*

8.1 Introduction

Water demand tends to exceed the available supply capacity, especially in many developing countries, as a result of rapid population growth (Ingeduld *et al.* 2006). As a consequence, Intermittent Water Supply (IWS) regimes are introduced with the aim of limiting that demand. Water scarcity in arid regions is also amplified when there is a lack of capacity in the distribution network, which is too deteriorated to deliver the required water (Ameyaw *et al.* 2013). Even though the water distribution should be equitable and efficient in such cases (Vairavamoorthy *et al.* 2008), intermittent supply have become a norm rather than an exemption (Seetharam and Bridges 2005), mainly due to necessity rather than the initial design (Vairavamoorthy *et al.* 2007).

IWS varies depending on the region and situation, ranging from few hours per day to few hours per week. Therefore, the consumers need to store as much water as possible during service hours, in order to compensate the periods of interruption. Often, this type of systems is characterized by inadequate pressures. Therefore, the water distribution is not homogenous (De Marchis *et al.* 2010), creating high peak factors in the distribution system (Andey and Kelkar 2009). Usually, in this type of regimes, pumps and pipes fail to carry the required water demand during the short periods required, leading to an unreliable service (Ameyaw *et al.* 2013). Furthermore, the low pressures derived from the high hydraulic losses, lead to increasing operational costs, higher leakage levels and higher consumers' costs. In addition, IWS may lead to the revenue losses, for it becomes more erratic and the willingness to pay for such a service declines (Ingeduld *et al.* 2006).

Individual storage facilities play important role in intermittent supply networks, since they are the only supply when the service is unavailable. The inflow into each household tank is based on the pressure conditions in the network and it equals to the maximum quantity of water that can be collected during the supply hours (Ingeduld *et al.* 2006). Therefore, the balance of water demand takes place for each individual household tank, "whereby replenishing of the volume behaves differently depending on water availability in the distribution network" (Trifunović and Abu-Madi 1999). Furthermore, the water is consumed according to the demand patterns that are not necessarily influenced by IWS regime.

Tourist islands face additional pressure to tackle the scarcity aiming to optimise the revenues from tourism while providing sufficient environmental protection. Puerto Ayora as the main tourist hub of Santa Cruz Island, has a distribution network built approximately 30 years ago, which consists of approximately 2,500 service connections and supplies brackish water to approximately 12,000 inhabitants, intermittently (INEC, 2010). Because of insufficient maintenance, the network is characterized by high leakage levels whose actual figure has not yet been established. Most households have storage facilities, mainly in the form of roof-tanks and/or cisterns, perceiving the supply as unreliable and insufficient. On the other hand, the fixed water-tariff structure seems to be the main cause of excessive water wastage within the premises. On top of it, the municipal supply service has not been able to cope with current tourist growth trends.

This chapter analyses the hydraulic performance of the current network in Puerto Ayora, with specific attention to the water losses from the spilling of the roof tanks. During the fieldwork data collection, a peculiar situation was observed regarding the local complaints on the lack of water combined with overwhelming amount of wastage from the overflow of roof-tanks. The origin seems to be in the absence of float valves and the fact that the owners do not close the faucets manually when the tank is already full (they may not be even at home at that moment). Therefore, an attempt was made to develop a suitable methodology for quantification of the overflows of household roof tanks and further reassess the currently applied IWS regime.

In the second part, the research aimed to assess the network conveyance capacity and the need for IWS and roof tanks, and propose suitable rehabilitation measures, for what has been indicated as the maximum demand growth scenario indicated in the research, described in previous chapters.

The hydraulic simulations for the first part of the study were conducted by running the Pressure-Driven Analysis (PDA), while the future demand scenario was simulated by the Demand-Driven Analysis (DDA).

8.2 Demand-Driven Analysis (DDA) and Pressure-Driven Analysis (PDA)

Network modelling softwares commonly use the Demand Driven Analysis (DDA) as the default hydraulic solver. This assumes that the nodal demands are known functions of time and are independent of the pressure available in the system (Cheung *et al.* 2005). The hydraulic solver then produces the nodal pressures and pipe/pump flows which satisfy those fixed nodal demands. The DDA usually presents reasonable and close-to reality solutions, but under regular supply conditions. This approach assumes that the nodal demands are always delivered.

Regardless of the pressures throughout the distribution system, the algorithm is able to formulate the needed equations in order to solve the unknown nodal heads (Ozger and Mays 2003). At the same time, this algorithm is unable to capture accurately the behaviour of intermittent systems, which operate under irregular conditions.

A number of studies, such as those by Germanopoulos (1985), Martinez *et al.* (1999), Soares *et al.* (2003), Hayuti and Burrows (2004) and Tanyimboh (2004), which were based mainly on field investigations, discuss the restrictions of DDA. These studies suggest the use of the PDA, which assumes a fixed demand above given pressure threshold, zero demand below the given minimum pressure, and proportional relationship between the pressure and the demand for the pressure range between the threshold and the minimum values (Cheung *et al.* 2005). The PDA approach uses the concept of orifices at system nodes, with the aim of replicating a pressure-demand relation in the modelling process. Therefore, the previous studies suggest that PDA can be more effective than DDA when simulating intermittent conditions.

More in detail, Tanyimboh *et al.* (2001) describe the pressure-driven demand relationship as:

$$H_i = H_i^{min} + K_i Q_i^n \qquad \text{(Equation 8.1)}$$

where H_i represents the actual head at demand node i, H_i^{min} refers to the minimum head to which below the service ends, K_i is the resistance coefficient for node i, Q_i refers to the nodal discharge flow, and n is the exponent that theoretically and usually takes the value of 2.0 (Gupta and Bhave 1996). Furthermore, if the value of Q_i is unknown for any given nodal head, then Equation 8.1 needs to be rearranged as follows:

$$Q_i = \left(\frac{H_i - H_i^{min}}{K_i} \right)^{1/n} \qquad \text{(Equation 8.2)}$$

If Q_i equals the required demand, Q_{req}, H_i should then equal the desired head in the node, named H_{des}. If the demand at that node further needs to be fully satisfied, then the head should be available as follows:

$$Q_i^{req} = \left(\frac{H_i^{des} - H_i^{min}}{K_i} \right)^{1/n} \Rightarrow \frac{1}{K_i^{1/n}} = \frac{Q_i^{req}}{\left(H_i^{des} - H_i^{min} \right)^{1/n}} \qquad \text{(Equation 8.3)}$$

Finally, Equation 8.4 is obtained by substituting K_i in Equation 8.2:

$$Q_i^{avl} = Q_i^{req}\left(\frac{H_i^{avl} - H_i^{min}}{H_i^{des} - H_i^{min}}\right)^{1/n}$$

(Equation 8.4)

Where, Q_i^{avl} refers to the flow for the head available at the node (H_i^{avl}). Equation 8.4 has three probable situations:

1) $H_i^{avl} \leq H_i^{min} \Rightarrow Q_i^{avl} = 0$

2) $H_i^{min} < H_i^{avl} < H_i^{des} \Rightarrow 0 < Q_i^{avl} < Q_i^{req}$

3) $H_i^{avl} \geq H_i^{des} \Rightarrow Q_i^{avl} = Q_i^{req}$

These situations are used when balancing the flows in the pipes, which are connected to node i. The key issue is the correct definition for H_i^{min} and H_i^{des}, such as their correlation with the nodal resistance K_i, which is the one that describes the nature of the PDD empirical relationship.

The EPANET software uses the PDA concept through Emitter Coefficients (EC), which models pressure-dependant flows from sprinkler heads, as developed by Rossman (2000). He describes the concept of EC by using similar relationships as in Equation 8.1. In this case, an emitter is modelled as a setup of a dummy pipe which is connected to the actual node, with a dummy reservoir whose nodal elevation (z) equals the initial head. Hence, $H_i^{min} = z_i$ and:

$$Q_i = \frac{1}{K_i^{1/n}}(H_i - z_i)^{1/n}$$

(Equation 8.5)

The K-value in Equation 8.5 refers to the resistance of the dummy pipe, but actually it has the same meaning as in Equations 8.1 to 8.3. Finally,

$$Q_i = k_i\left(\frac{p_i}{\rho g}\right)^{\alpha} \quad ; \quad \frac{p_i}{\rho g} = H_i - z_i \quad ; \quad \alpha = 1/n \quad ; \quad k_i = \frac{1}{K_i^{\alpha}}$$

(Equation 8.6)

Where k_i is the EC in node i and α is an emitter exponent with theoretical value of 0.5. EC was first introduced to simulate operation of fire hydrants.

Several other PDA approaches have been based on further improvement of the EC concept, such as the one by Pathirana (2010). In order to use the PDA with sufficient degree of accuracy, extensive field data collection is required for determination of the relationship between nodal heads and flows (Ozger and Mays 2003).

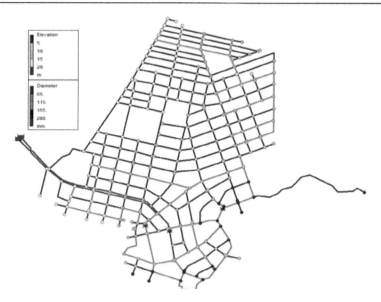

Figure 8.1- Layout of Puerto Ayora water supply network

Figure 8.2- Puerto Ayora's network with roof tanks

8.3 Research methodology

The gravity hydraulic network model was built in the EPANET software, as shown in Figure 8.1, which consists of 2 reservoirs, 284 nodes and 367 pipes. An estimation of nodal demands was done based on the total population and the demography per m², using the demand patterns established from the field data (Chapter 5).

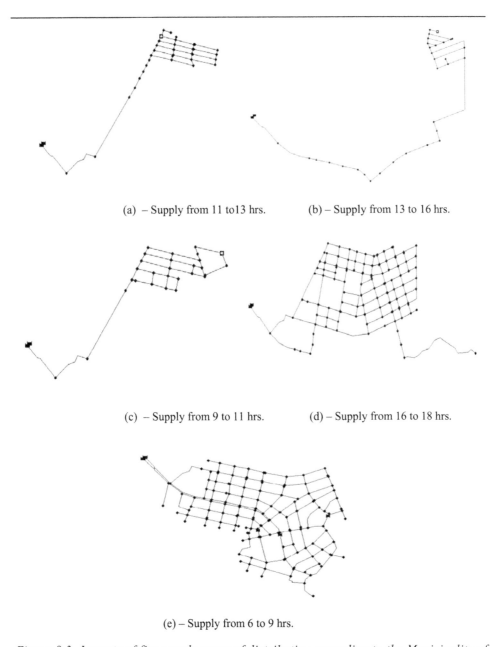

(a) – Supply from 11 to13 hrs.   (b) – Supply from 13 to 16 hrs.

(c) – Supply from 9 to 11 hrs.   (d) – Supply from 16 to 18 hrs.

(e) – Supply from 6 to 9 hrs.

Figure 8.3- Layouts of five supply zones of distribution according to the Municipality of Santa Cruz.

To check the robustness and numerical stability of the EPANET hydraulic solver, a variant was also developed by modelling each node connected to a tank of the size anticipated from the number of occupants supplied from, and its elevation corresponding to that of the node; this model is shown in Figure 8.2. An average of five inhabitants per household was assumed and an average tank height of 1.5 m above the ground level. All tanks were assigned the initial depth of 1 m, the minimum depth = 0 m, and the maximum depth = 3 m. Also, a check-valve was modelled on a short dummy pipe, to prevent backflow from the tank (i.e simulate the inlet arrangement from the top).

Alternatively, the PDA approach was simplified in order to enable the rationing as applied in reality. To make it possible, the network with emitter coefficients was divided into five small sub-systems (Figure 8.3), capturing the five distribution zones created by the municipality, shown in Chapter 5 (Figure 5.1). All peak factors are shown in Table 8.1. Different EC values were tried for each zone and supply hours, in an attempt to match the previously reported total amount of supply per day (which is approx. 5000 m³ for the horizon 2014, including the leakage), ending with the adopted EC value of 0.5.

The assessment of the overflow of nodal tanks was done per zone, for several scenarios created by varying the leakage levels, the average volume of individual storage, and the percentage of the level of the tank filled at the moment the water supply starts in a particular distribution zone.

The volumes considered in making the balance per each node of the five sub-models are further shown in Equations 8.7 – 8.10.

$$V_i^{avl} = n_{h,i} X_1 \qquad\qquad \text{(Equation 8.7)}$$

V_i^{avl} is the total available individual storage volume in node i (in m³), $n_{h,i}$ is the number of households served from the node, and X_1 is the variable that indicates the average storage volume available per household (m³).

$$V_i^{c24} = n_{c,i} X_2 \qquad\qquad \text{(Equation 8.8)}$$

V_i^{c24} is the total volume consumed in node i over 24 hours (in m³), $n_{c,i}$ is the number of consumers served from the node, and X_2 is the variable that indicates the average specific demand per capita (lpcpd).

$$V_i^{s24} = Q_i^{EC} h_i^s \left(1 - \frac{X_3}{100}\right) \qquad\qquad \text{(Equation 8.9)}$$

V_i^{s24} is the total volume supplied to node i over 24 hours (in m³); this is an IWS that occurs during h_i^s hours at the flow Q_i^{EC} (in m³/h) based on the available pressure calculated in EPANET using the emitter coefficients. X_3 is the variable that indicates the average leakage percentage in node i.

$$V_i^{cIWS} = \sum_{j=1}^{h_i^s} \frac{V_i^{c24}}{24} pf_{j,i}$$
(Equation 8.10)

V_i^{cIWS} is the total volume consumed in node i (in m³) during the IWS period of h_i^s hours, at hourly peak factors $pf_{j,i}$ applied depending on the period of the day when the IWS takes place (for the diurnal patterns, see Chapter 5). Hence, the actual volume accumulated in the tank(s) of node i during the IWS period is $V_i^{s24} - V_i^{cIWS}$. Assuming X_4 to be the variable that indicates the percentage of the total available volume V_i^{avl} already occupied at the moment when the IWS starts, the buffer of volume in the tank(s) of node i, V_i^{buf}, when the IWS stops, will be:

$$V_i^{buf} = V_i^{avl} \left(1 - \frac{X_4}{100}\right) - \left(V_i^{s24} - V_i^{cIWS}\right)$$
(Equation 8.11)

Possible negative result in Equation 8.11 will indicate the overflow i.e. the spilling from the tank(s) of node i. Furthermore, when the IWS stops, the tank(s) will be discharged for the volume $V_i^{c24} - V_i^{cIWS}$ suggesting the initial volume before the IWC starts again to be:

1) $V_i^{avl} - (V_i^{c24} - V_i^{cIWS})$, if the overflow was taking place during the IWS ($V_i^{buf} < 0$);

2) $V_i^{avl} - V_i^{buf} - (V_i^{c24} - V_i^{cIWS})$, if the overflow was <u>not</u> taking place during the IWS ($V_i^{buf} \geq 0$).

In both of these cases, the result can in theory be negative, suggesting the water shortage and/or insufficient volume of the tanks. Assuming that this is not the case, some indication exists while assessing the values for X_4. Sample calculation done for IWS Zone 1 shown in Figure 8.3(a) is given in Table 8.2, taking $X_1 = 1.5$ m³, $X_2 = 163$ lpcpd, $h_i^s = 2$ hours, $X_3 = 17.5\%$, $pf_1 = 1.29$, $pf_2 = 1.66$, and $X_4 = 0$. With these, the total overflow amounts at approximately 114 m³, which is about 42% of the total volume supplied (of 269 m³).

Table 8.1-Demand multipliers used for Puerto Ayora water distribution network

Hour	1	2	3	4	5	6	7	8	9	10	11	12
pf	0.53	0.53	0.53	0.53	0.53	2.22	1.14	1.20	1.24	1.28	1.29	1.66
Hour	13	14	15	16	17	18	19	20	21	22	23	24
pf	1.18	1.18	1.09	1.04	1.35	1.63	1.69	0.53	0.53	0.53	0.53	0.53

ID_i	$n_{c,i}$	$n_{h,i}$	V_i^{tank} (m³)	X_2 (lpcpd)	h_i^s (hours)	V_i^{tc24} (m³)	V_1^{clWS} (litres)	V_2^{clWS} (litres)	Q_i^{EC} (l/s)	V_i^{tc24} (m³)	$V_i^{tc24} - V_i^{clWS}$ (m³)	V^{buf} (m³)
J-1	31	6	6.2			5.1	272.9	351.2	2.1	12.4	11.7	-5.5
J-2	29	6	5.8			4.7	252.1	324.4	2.0	12.1	11.5	-5.8
J-3	18	4	3.6			2.9	157.0	202.1	2.1	12.5	12.2	-8.6
J-42	15	3	3.0			2.5	133.1	171.3	2.1	12.4	12.1	-9.1
J-43	27	5	5.4			4.4	238.1	306.4	2.0	11.8	11.3	-5.8
J-44	15	3	3.0			2.5	132.2	170.1	2.0	11.8	11.5	-8.5
J-45	27	5	5.3			4.3	233.3	300.2	2.1	12.5	11.9	-6.6
J-46	39	8	7.8			6.4	342.1	440.3	2.0	11.9	11.2	-3.4
J-47	23	5	4.5			3.7	198.8	255.8	2.1	12.6	12.1	-7.6
J-48	47	9	9.3			7.6	408.9	526.2	2.0	11.6	10.7	-1.3
J-50	27	5	5.4	163		4.4	238.3	306.7	1.8	10.5	10.0	-4.5
J-64	28	6	5.7		2	4.6	249.4	321.0	2.0	11.7	11.1	-5.4
J-65	54	11	10.9			8.8	475.7	612.1	2.1	12.5	11.4	-0.5
J-66	27	5	5.4			4.4	237.6	305.8	2.1	12.5	11.9	-6.5
J-67	44	9	8.9			7.2	389.5	501.3	2.0	12.1	11.2	-2.3
J-68	28	6	5.6			4.5	244.3	314.3	2.2	12.8	12.3	-6.7
J-69	35	7	7.1			5.8	311.0	400.2	2.2	13.2	12.5	-5.4
J-70	29	6	5.8			4.7	254.1	326.9	2.0	12.1	11.5	-5.7
J-78	54	11	10.8			8.8	473.6	609.4	2.1	12.7	11.6	-0.8
J-79	47	9	9.3			7.6	407.9	524.9	2.0	11.8	10.9	-1.6
J-80	26	5	5.2			4.2	227.6	292.9	2.0	11.7	11.2	-6.0
J-81	34	7	6.8			5.5	298.0	383.5	2.3	13.5	12.9	-6.1

Table 8.2-Assessment of tanks' overflow in IWS Zone 1

Table 8.3 -Local and tourism population calculation for the future 30 years in Puerto Ayora

YEAR	Total growth rate	Local population (inhab.)	Demand local population (m³/day)	No. of tourists/ year	Tourist consumption (m³/day)
SLOW GROWTH					
2015	-	15,801	2,576	205,505	1,109
2020	0.05	16,607	2,707	215,780	1,165
2025	0.10	17,454	2,845	226,055	1,220
2030	0.15	18,345	2,990	236,331	1,276
2035	0.20	19,280	3,143	246,606	1,331
2040	0.25	20,264	3,303	256,881	1,386
2045	0.30	21,087	3,437	267,156	1,442
MODERATE GROWTH					
2015	-	15,801	2,576	205,505	1,109
2020	0.15	18,318	2,986	236,331	1,276
2025	0.30	21,235	3,461	267,156	1,442
2030	0.45	24,618	4,013	297,982	1,608
2035	0.60	28,539	4,652	328,808	1,775
2040	0.75	33,084	5,393	359,634	1,941
2045	0.90	37,236	6,070	390,459	2,107
FAST GROWTH					
2015	-	15,801	2,576	205,505	1,109
2020	0.25	20,167	3,287	256,881	1,386
2025	0.50	25,738	4,195	308,257	1,664
2030	0.75	32,849	5,354	359,634	1,941
2035	1.00	41,925	6,834	411,010	2,218
2040	1.25	53,508	8,722	462,386	2,496
2045	1.50	65,040	10,601	513,762	2,773
VERY FAST GROWTH					
2015	-	15,801	2,576	205,505	1,109
2020	0.35	22,174	3,614	277,432	1,497
2025	0.70	31,116	5,072	349,358	1,886
2030	1.05	43,665	7,117	421,285	2,274
2035	1.40	61,275	9,988	493,212	2,662
2040	1.75	85,987	14,016	565,139	3,050
2045	2.10	112,759	18,380	637,065	3,438

The full analysis was done for the entire network, and then the influence of each variable was judged. Finally, the water supply network was assessed for the future under the same four tourist and local population growth scenarios used in Chapter 6 (Mena *et al.*, 2013). The analysis was done in the intervals of five years, as shown in Table 8.3. The same growth rate was used for all nodal demands, while the tourist demand was distributed amongst the nodes located in the centre of the town where most of the tourists reside.

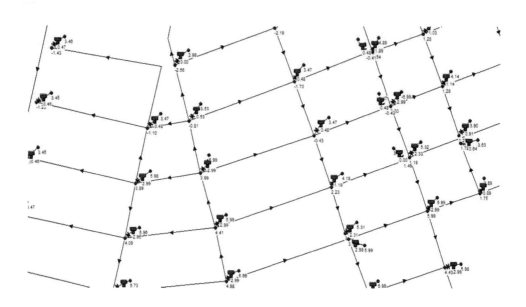

Figure 8.4- Sample network layout with nodal tanks

8.4 Simulations, results and discussion

8.4.1 Current situation simulated with nodal tanks

The fist simulation runs were conducted with the network model shown in Figure 8.2. The statistics of this model indicates 578 nodes, 286 tanks and 940 pipes. Significantly larger number of pipes originates from the setup of the discharge from the tanks that consists of a dummy pipe and the node with the attached consumption pattern, and another dummy pipe on the upstream side of each tank, which is a check valve simulating the inlet arrangement from the top; once the water is in the tank, no water can go back into the network. In this way the model becomes rather bulky, but this is the only way to reflect 100% accurately the actual reality.

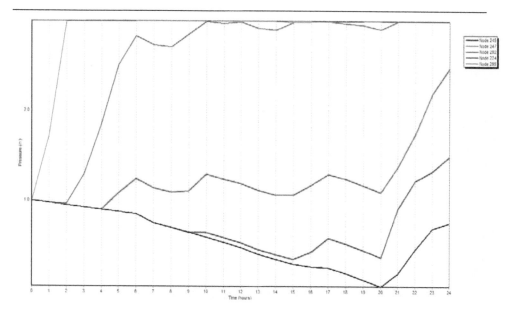

Figure 8.5- Pressures over 24 hours for randomly selected tanks

The simulation runs did not yield any tangible result. First of all, the model appeared to be very sensitive to the selection of the tank properties and the initial water levels. As soon as the tank gets emptied or filled from the imbalance between the demand and supply, based on the EPANET settings, the tank becomes disconnected from the network. This suggests that the multiple occurrences inflicts numerical instability, which was also experienced by significantly longer simulation times (although not dramatically long), and resulting in negative pressures in the network.

Figure 8.4 shows the detail from the network, documenting the negative pressures, while Figure 8.5 shows the trends of water level variation in randomly selected tanks in the same area. Hence, the calibration of such a model appeared to be very complex and it seems that with this amount of tanks, the model could work only if these are not entirely emptied or completely full. The model becomes unstable and suggests that it might work if all the tanks will have sufficient amount of water throughout the entire day. In that case, analyses of any irregular scenario becomes practically impossible, which was the main reason to abandon this way of modelling.

Table 8.4- Scenarios for the estimate of tank overflows in Puerto Ayora

X_2 (lpcpd)	X_3 (%)	X_1 (m³)	X_4 (%)	V_l^{buf} (m³)	V_l^{buf} (2) (m³)	% total daily supply
				Scenario 1		
	17.5	1	0	-1327	-929	18.6
				Scenario 2		
	17.5	1	50	-2301	-1611	32.2
				Scenario 3		
	17.5	2	0	-423	-296	5.9
				Scenario 4		
163	17.5	2	50	-709	-496	9.9
				Scenario 5		
	25	1	0	-1053	-737	14.7
				Scenario 6		
	25	1	50	-1957	-1370	27.4
				Scenario 7		
	25	2	0	-287	-201	4.0
				Scenario 8		
	25	2	50	-1053	-737	14.7
				Scenario 9		
	17.5	1	0	-1285	-900	18.0
				Scenario 10		
	17.5	1	50	-2239	-1567	31.3
				Scenario 11		
	17.5	2	0	-411	-287	5.7
				Scenario 12		
195	17.5	2	50	-1285	-900	18.0
				Scenario 13		
	25	1	0	-1016	-712	14.2
				Scenario 14		
	25	1	50	-1900	-1330	26.6
				Scenario 15		
	25	2	0	-298	-208	4.2
				Scenario 16		
	25	2	50	-1016	-712	14.2

8.4.2 Results of Pressure-Driven Approach with Emitter Coefficients

The alternative way of modelling was therefore adopted as elaborated by Equations 8.7 to 8.11 and in Table 8.2. The 16 selected scenarios based on the values X_1 to X_4 and the results showing the buffer volume are to be seen in Table 8.4. Two specific demand scenarios (X_2) correspond to the figures obtained by the field survey (Chapter 4) and earlier pilot measurements done by the municipality (Chapter 5). Similarly, the leakage scenarios (X_3) were determined from the research in previous chapters assuming a NRW of 35%. Therefore, the physical leakage was assumed 50% of the total NRW (a minimum figure-17.5%), and a figure between total NRW and minimum assumed leakage (25%). The negative buffer indicates the spilling from the tanks with the 2nd figure showing the value reduced for 30%, which is believed to be the percentage of the households having the float valves installed on their roof tanks.

Table 8.5- Additional population to be supplied based on different overflow scenarios

Scenarios	Specific Demand of 163 lpcpd	Percentage of population (%)	Specific Demand of 195 lpcpd	Percentage of population
1	5697	33.5	4762	28.0
2	9882	58.1	8261	48.6
3	1817	10.7	1519	8.9
4	3043	17.9	2543	15.0
5	4521	26.6	3779	22.2
6	8403	49.4	7024	41.3
7	1234	7.3	1032	6.1
8	4521	26.6	3779	22.2
9	5519	32.5	4614	27.1
10	9614	56.6	8036	47.3
11	1763	10.4	1474	8.7
12	5519	32.5	4614	27.1
13	4365	25.7	3649	21.5
14	8158	48.0	6819	40.1
15	1278	7.5	1068	6.3
16	4365	25.7	3649	21.5

In terms of the percentage of daily supply, the results show relatively wide range: between 4 and 31.3 %, which indicates sensitive input data. Based on the field observations, it is believed that Scenario 6 may be the closest to the reality, with remark that the tank levels at the beginning of the IWS next day may well have different filling percentages, based on the consumption pattern of the previous day; hence, the values of X_4 need specific validation compared to the other three variables.

Based on the results in Table 8.4, an assessment of the additional supply in case the overflow from the tanks could have been prevented, was done. Table 8.5 shows additional number of consumers that could be supplied for each of the 16 scenarios, assuming two specific demands used in the analysis of the tank buffers. As can be seen, in the most extreme situation of Scenario 2, almost 60% of extra population could have been supplied at the lower specific demand of 163 lpcpd, and nearly 50% at the higher one of 195 lpcpd. If true, this would put a valid hypothesis about the necessity of the roof tanks for they may not ease, but actually boost the intermittency resulting from the negligence of local population.

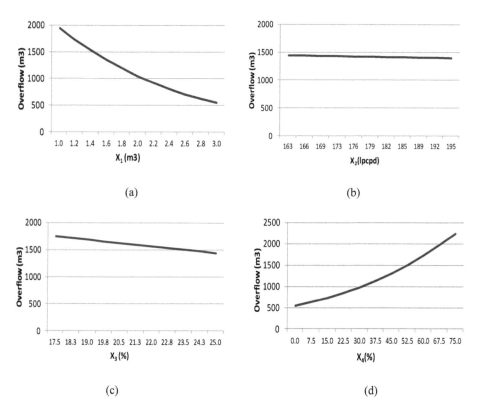

(a)

(b)

(c)

(d)

Figure 8.6- Tank overflow based on different (a) individual tank volume, (b) specific demand, (c) leakage level and (d) tank level in total overflow estimation.

The trend of sensitivity of the four X-variables was further analysed against the tank overflow in the somewhat wider range than the one applied in the 16 scenarios. Figure 8.6 shows that the highest sensitivity belong to the variables X_1 and X_4, which underpins the points on further fieldwork to validate the available volume (individual household storage) and the initial level of the tanks before the IWS kicks-off, every day. On the other hand, the specific demand, as well as the average leakage percentage, influences the overflow volume to a lesser extent.

8.4.3 Demand-Driven Analysis for future growth modelling

The analyses of the future demand growth, as well as the rehabilitation measures needed were done by applying the DDA. The results obtained from the demand scenarios shown in Table 8.3 are given in Table 8.6. This table shows the percentage of the nodes in three different pressure ranges (less than 0, 0 to 10, and greater than 10 mwc) for the daily demand scenario at 6:00 hrs, which is assumed as the maximum consumption hour.

Not surprisingly, it can be observed that the supply at the year 2045 for the very fast growth is practically impossible. Actually, both the fast and very fast growth rates are out of proportion regarding the size and capacity of the current network; the simulations show that already at year 2020, the network will start to perform poorly. Therefore, a major and very expensive renovation would need to be considered, resulting in extremely high financial burden for the municipality. Here, one would also need to consider a very long period of construction with permanent disturbances to the tourists and local population.

Only in the slow population growth scenario the network would perform optimally for the next 30 years. The calculations showed that the nodal pressures would hardly drop below 20 mwc at the peak hour, suggesting no rehabilitation of the current network would be necessary, under this population growth scenario.

However, the research results reported in Chapter 7, put the moderate growth scenario (3% annual growth for local population and 4% annual growth for tourist visitors) as the preferred one by the majority of stakeholders, and the maximum recommended one in order to cover water demand with supply. Further network analysis was therefore done taking into consideration this scenario. Two options were considered in order to solve the water deficit and lack of pressures in the moderate growth scenario at year 2045.

Table 8.6- Percentage of nodes with different ranges of pressure at peak hour (6:00 hrs)

Year	% of nodal pressures		
	< 0 mwc	0 - 10 mwc	> 10 mwc
VERY FAST GROWTH			
2015	0%	0%	100%
2020	5%	7%	87%
2025	85%	1%	14%
2030	86%	1%	13%
2035	86%	1%	13%
2040	90%	2%	8%
2045	93%	3%	4%
FAST GROWTH			
2015	0%	0%	100%
2020	0%	0%	100%
2025	0%	10%	90%
2030	48%	24%	28%
2035	75%	8%	17%
2040	84%	2%	14%
2045	86%	0%	14%
MODERATE GROWTH			
2015	0%	0%	100%
2020	0%	0%	100%
2025	0%	3%	97%
2030	0%	7%	93%
2035	3%	56%	41%
2040	53%	20%	27%
2045	72%	9%	19%
SLOW GROWTH			
2045	0%	0%	100%

The first one included a partial renovation of the network, increasing some of the pipes' capacity, assuming there would be enough water to be extracted for distribution. For this analysis, an average consumption of 163 lpcpd was assumed also for the future condition following the intervention strategy proposed in Chapter 6 to reduce the water demand; thus, not increase the specific demand. The average is assumed to be constant throughout the year, with no major seasonal variations. An intervention to guarantee the minimum pressure of 25 mwc is shown in Table 8.7, knowing that the authorities want to target elite tourism, i.e. better level of service should be provided, especially for the hotels.

Table 8.7- Pipes to be changed in the moderate growth scenario for minimum pressure of 25 mwc at peak hour (in 2045)

ID	Node 1	Node 2	Length (m)	Initial Diameter (mm)	Changed Diameter (mm)	Approximate cost (EUR)
P-194	J-125	J-159	466.8	200	250	38,883
P-195	J-125	J-159	466.8	160	250	38,883
P-196	J-159	J-160	271.0	200	250	22,576
P-277	J-223	J-224	94.5	200	250	7,871
P-279	J-226	J-223	69.7	200	250	5,808
P-294	J-234	J-236	70.0	200	250	5,835
P-295	J-236	J-225	102.4	200	250	8,528
P-302	J-238	J-241	77.0	200	250	6,417
P-303	J-241	J-237	61.1	200	250	5,088
P-386	J-310	J-311	68.1	110	150	4,519
P-389	J-312	J-314	45.6	110	150	3,027
P-390	J-314	J-281	34.6	63	150	2,298
P-392	J-315	J-316	10.8	63	150	720
P-427	J-341	J-343	51.9	63	150	3,448
P-443	J-276	824-A	15.4	200	250	1,281
P-445	J-314	827-A	13.1	200	250	1,095
P-447	J-267	J-272	83.6	110	150	5,551
P-464	J-238	J-245	74.3	200	350	10,155
P-465	R-3	J-160	267.9	200	300	27,010
TOTAL						**198,994**

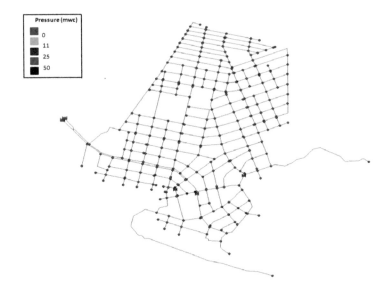

Figure 8.7- Pressures with changing of pipes for a minimum of 25 mwc with moderate growth scenario at year 2045.

In order to satisfy the demand at the end of the planning horizon for the moderate growth scenario, the approximate investment adds up to nearly 200,000 EUR, which is somewhat unexpected, since it does not seem a significant amount in view of the population growth rate. However, this cost does not include yet the high excavation costs of volcanic rock, and therefore, the cost will increase considerably. Despite this increase, there would be enough time for the Municipality of Santa Cruz to raise the funds; for instance, in the form of additional tax paid by the tourists. In this scenario, 19 pipes are to be renovated and the effect on the pressures is to be seen in Figure 8.7.

Table 8.8-Analysis of balancing volume for the maximum consumption day in 2045, for moderate population growth scenario.

Hour	Tank In (m³/h)	Tank Out (m³/h)	Peak Factor	In-Out (m³/h)	Cumulative Volume (m³/h)
1	322.6	167.5	0.5	155.1	155.1
2	322.6	167.5	0.5	155.1	310.2
3	322.6	167.5	0.5	155.1	465.3
4	322.6	167.5	0.5	155.1	620.4
5	322.6	167.5	0.5	155.1	775.5
6	322.6	701.5	2.2	-378.9	396.6
7	322.6	360.3	1.1	-37.7	358.9
8	322.6	379.2	1.2	-56.6	302.3
9	322.6	391.9	1.2	-69.3	233.0
10	322.6	404.5	1.3	-81.9	151.1
11	322.6	407.7	1.3	-85.1	66.0
12	322.6	524.6	1.6	-202.0	-136.0
13	322.6	372.9	1.2	-50.3	-186.3
14	322.6	372.9	1.2	-25.0	-236.7
15	322.6	347.6	1.1	-6.1	-261.7
16	322.6	328.7	1.0	-104.0	-267.8
17	322.6	426.6	1.3	-192.5	-371.8
18	322.6	515.1	1.6	-211.5	-564.3
19	322.6	534.1	1.7	155.1	-775.8
20	322.6	167.5	0.5	155.1	-620.7
21	322.6	167.5	0.5	155.1	-465.6
22	322.6	167.5	0.5	155.1	-31.5
23	322.6	167.5	0.5	155.1	-155.4
24	322.6	167.5	0.5	155.1	-0.3
Balancing volume (Max. Cum. Vol + ABS. Min. Cum. Vol.)					775.5 +775.8= 1551.4

The previous analysis assumes that the roof tanks would become redundant in the future, to larger extent for the sake of water quality risks, which was already reported by Liu (2011). Alternatively, in the second option these tanks were still considered, because they traditionally have a high public acceptance. Moreover, the tanks also have a positive hydraulic impact into the renovation of the distribution network, which could be then based on the average flow rather than the peak hourly flows.

In the second option using the moderate growth scenario, an average flow in the network of 121 l/s in the year 2045 was assumed, reducing 35% of leakage (*pf* value closer to 1). For this value, Table 8.8 shows the balancing volume on the maximum consumption day and Figure 8.8 shows the demand patterns, as well as the total balancing volume percentage.

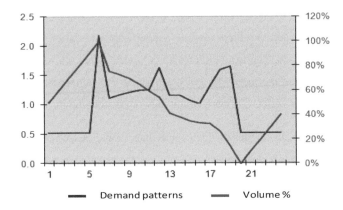

Figure 8.8- Demand patterns in Puerto Ayora and total balancing volume percentage

As observed in Table 8.8, this steady state calculation results in balancing volume of 1,551.4 m³. If it is assumed that the population in year 2045 will be 37,236 inhabitants (), and an average occupation of 5 inhabitants, there would be a total of 7,447 households, resulting in average individual storage tanks of 0.21 m³. This option means that the investment costs of the tanks should be taken into consideration. These costs could be assumed either by the owners of each property or by the municipality. Alternatively, a central service tank(s) could also be constructed due to a favourable topography of the area, but placing it in relatively densely populated (tourist) area may pose a problem, also from an aesthetical point of view.

8.5 Conclusions

Three modelling approaches are illustrated in this chapter, including nodal tanks, the PDA and the DDA, for the current and future demand in Puerto Ayora. In the first case, it was very complicated to replicate the actual individual storage in the model. The EPANET software becomes unstable due to different water level patterns in each tank, sometimes full and sometimes empty. More detailed information would be needed for a proper calibration of such a model but even then it is possibly that the model could perform only regular demand scenarios. Unfortunately, EPANET disconnects empty tanks which distorts its numerical stability in multiple occurrence of this situation.

The PDA of present demand reflects partially the situation with the household storages. The aim was to estimate the overflow of roof tanks under several scenarios. As observed from the results, it can be concluded that there may be a significant amount of water lost due to the overflow of the roof tanks. The local authorities in Puerto Ayora, as well as local residents, have the erroneous idea of the "need" for individual storage to compensate the lack of water in the hours of no supply, which to the large extent may result from the negligence.

The trend of the sensitivity suggests which input data should be verified with priority. The results show the size of individual household tanks and the initial level when the supply begins have more impact on the overflow, than the leakage percentage or per capita demand. Despite the fact that many assumptions were made, this analysis provides a practical approach to measure the volume that might be spilled from household tanks, which seems to be the main source of water wastage. The municipality would need to monitor and record individual characteristics of the households' storage facilities, in order to assess more accurately the extent of this problem. Also, the model should be further calibrated by adequate choice of emitter coefficients.

Regarding the future assessment, in the case of very fast and fast growth scenario, a major renovation of the network will be needed. Therefore, the moderate growth scenario would be an adequate one. For this scenario, a gradual network renovation is proposed, considering the replacement of main pipes. With this option, the roof tanks could be removed. On the other hand, by calculating the gross balancing volume on the maximum consumption day, the roof tanks will be needed to cover the peak hour supply, but the IWS regime could be avoided. This further means no major investment into the network renovation but into the storage provision, be it in the form of individual household tanks or a centralised storage in the town. The size of the future tanks seems relatively smaller in comparison with the current average tank size; which hints that present household tanks have bigger dimensions than it is necessary and as a result the demand is also higher.

Finally, the research points that the hydraulic modelling of the distribution network of Puerto Ayora poses quite a complex problem due to: (1) numerical instability caused by multiple tanks existing in the model, and (2) difficult calibration from lots of unknown and inaccurate data needed to build a reliable model.

"Science never solves a problem without creating ten more".

-George Bernard Shaw

9

GENERAL CONCLUSIONS AND FUTURE OUTLOOK

The water resources of Santa Cruz Island, a tourist hub of the Galápagos Archipelago, are severely threatened by growth in the local population and the number of tourists visiting the island. Tourism has increased from approximately 18,000 visits in 1980 to more than 220,000 in 2015. As a result, water resources are being overexploited to try to meet the needs of tourists and related businesses. The island municipality has not been able to cope with water demands resulting from this growth, providing inadequate water supplies. In addition, limited resource and population data contributes to the lack of practical solutions to effectively address water scarcity on Santa Cruz.

The overall aim of this research was to assess water supply and demand patterns on Santa Cruz, in order to generate sufficient data to develop sustainable management solutions that can overcome future water scarcity.

9.1 Water supply and demand on the Island of Santa Cruz

The lack of freshwater resources on Santa Cruz has been a problem for the local population for a number of years. Since there are no permanent sources of freshwater on the island, the municipality is only able to supply (untreated) brackish water, which is pumped from crevices of a basal aquifer. This brackish water is used for most household activities other than drinking. Island consumers buy bottled drinking water that is produced by small, brackish water desalination plants on the island. This has been the situation for residents and tourists since the municipal water supply began operation in the 1980s.

Santa Cruz continues to deal with the effects of a rapidly increasing tourist industry, like other islands around the world. Water supply problems, caused mainly by the exponential local population and tourism growth over the last two decades, include an unreliable supply, poor brackish water quality (chloride levels from 800 to 1200 mg/L), fixed water tariff structures, lack of water metering, and high water losses in the distribution system (ranging from 35% to 70%, according to different studies). These factors have led to a deeply rooted – albeit not completely accurate – perception of chronic water shortage by locals and authorities.

It is correct that freshwater resources are scarce on Santa Cruz. Based on the results of Chapter 3, the total supply is approximately enough to supply ±371 litres per capita per day (lpcpd), with the municipal supply being ±226 lpcpd. Leakage levels were not considered at this point. As indicated in Chapter 4, the water demand of domestic users, was on average 200 lpcpd. The results from a survey conducted on 388 premises on the island suggested that the average municipal water consumption in the town of Puerto Ayora is ± 163 lpcpd, and ± 96 lpcpd in Bellavista. Additionally, when the other types of water sources were also considered (bottled-desalinated water and water extracted from private crevices), the figures increased to ± 177 lpcpd and ± 253 lpcpd, respectively. These average per capita consumptions are similar to values in mainland Ecuador (160 lpcpd in Quito), where water resources are abundant. Therefore, the average specific demand estimated for Santa Cruz is very high for this type of insular and water scarce area. These results also indicate that the major consumers are hotels, accounting for almost 55% of total water demand and for 30% of the municipal water demand.

Results from Chapter 5 point to a wide range of domestic consumption patterns, based on different information sources used for estimating the per capita demand. Firstly, information from three metered pilot zones was analysed, and results revealed an average water demand ranging from 182 to as high as 428 lpcpd. Moreover, based on the measurements of 18 water meters installed in Puerto Ayora for the purpose of this research only, the domestic demand ranged from 52 to 420 lpcpd. These extremely high consumption amounts contradict the perception of a water shortage, and further indicate that the current brackish water resource is accepted by locals for uses other than drinking. This high water consumption also reveals a lack of proper monitoring of informal tourist accommodations by the Ministry of Tourism, since the higher averages seem unrealistic for domestic categories.

With the use of Demand-Driven Analysis and Pressure-Driven Analysis (PDA) approaches in Chapter 8, it was suggested that the system could currently provide a 24-hour water supply service, based on the measured pressures and other hydraulic characteristics within the distribution network. These approaches also provided a clear picture in which storage facilities are negatively affecting the flow of water supply instead of easing the situation. Moreover, this problem is worsened by the lack of stop-valves in roof tanks – resulting in overflow of the tanks, as well as the lack of monitoring of such daily events. Overflowing of roof tanks is common and contributes significantly to the irregularity of water supply. This excessive wastage from storage facilities is most likely due to the low,fixed water tariff structure (5.25 USD/month), which is seen as the main reason why the local population wastes water, along with the poor quality of the supplied brackish water. The municipal authorities have not raised water tariffs, purportedly because of fear of the reaction of the consumers, especially since the water is not suitable for drinking. In addition, political reasons contribute to the current water tariff structure, and this is a factor in the lack of awareness of water resource conservation.

Based on the analysis of the survey, the assessment of water meter data and the modelling of network performance, the more serious water problems are more related to poor water quality, rather than low quantity. In conclusion, in terms of the available supply capacity, as well as the water distribution infrastructure, there is currently no evidence to support the serious need for an intermittent supply system. Therefore, complaints regarding the current lack of water are evidently unfounded since the brackish groundwater is sufficient for most needs of the local population. The total supply of 370 lpcpd is about twice the average supply of a resident in Quito. However, water quality remains the major issue, which is somewhat overcome with bottled-desalinated water from local desalination companies.

9.2 Future water supply and demand management options

Based on the current water supply situation and elevated water demand (5220 m³/day), different strategies were proposed and analysed with the Watermet² software, in order to solve future water shortages. Firstly, individual alternatives, such as leakage reduction, water meter installation, rainwater harvesting, greywater recycling, demand reduction and the installation of a seawater desalination plant, were considered in Chapter 6. However, after modelling these alternatives individually, only the installation of a seawater desalination plant met projected demand up to 30 years. For this reason, these individual alternatives were combined into five complementary intervention strategies. They were also evaluated based on the coverage of water demand with supply over the next 30 years, under four growth scenarios (Chapter 6). Even with this combination, none of the proposed intervention strategies reached full coverage of water demand for categories of population growth scenarios: 1) very fast 2) fast 3) moderate, except the strategy that included a seawater desalination plant (Strategy # 2). Importantly, this strategy is the only option that significantly improves the water quality. On the other hand, Strategy # 5 (which considered all of the alternatives except desalination), was able to reach 100% of the water demand, but only for a slow scenario of population growth, and 70% of demand in the moderate growth scenario. In both of these scenarios, the use of bottled drinking water would need to continue. Even though the slow growth scenario has been recommended by environmentalists and NGOs, it is a highly unrealistic one, since the tourism industry continues to expand at a rate of 7% annually. On the other hand, the moderate scenario seems a more viable option, whilst the fast and very fast growth scenarios (preferred targets by the government) would worsen water supply and quality issues. With the government targets for tourism levels, seawater desalination appears to be the only viable option to fully meet demand with a higher water quality. However, environmental concerns of desalination related to brine disposal and other discharges, as well as high energy consumption, are important considerations in this fragile ecosystem. Despite the fact that a brackish water desalination plant could be installed instead, it was not considered as an option in order to protect the basal aquifer. Even though there is no major research on the levels of sea water intrusion and mixing rates in the intrusion zone, as well as on the change of salinity of brackish water over time, these potential problems need to be taken into account. A study made by Pryet (2013), suggested that the recharging rate for the aquifer where La Camiseta crevice is installed, is slow. Therefore, if water extraction increases significantly, this will most likely impact the chloride levels and total dissolved solids over the next 30 years and, eventually, the installation of a seawater desalination plant will be inevitable.

Chapter 7 also suggested that the moderate population growth scenario (3% annually for locals and 4% for tourists) is the preferred scenario for the island. Even though this scenario can be considered high in comparison to growth rates of other tropical islands, sustainable options can still be applied to avoid environmental impacts from the installation of a seawater desalination plant. According to the Multi-Criteria Decision Analysis (MCDA) carried out in Chapter 7, based on feedback from different stakeholders, the preferred intervention strategy is # 5 (combination of rainwater harvesting, greywater recycling, demand reduction, installation of water meters and leakage reduction) for local experts, domestic users and hotels. However, according to the decision-makers group, strategy #2 is preferred (including seawater desalination), because it solves water quality issues as well. This suggests that even though Strategy # 5 does not significantly improve water quality, the vulnerability of the ecosystem is considered by the main consumers. This is difficult to ignore since the main attraction of the island is its unique natural beauty and biodiversity. Any controversy over the preferences also illustrates the complexity of the dilemma. On one hand there is a pristine ecosystem to be preserved, and on the other hand, basic services such as potable water need to be provided. The development of relevant and specific policies for Santa Cruz, demands a deeper analysis of the four main selected criteria used for the MCDA (environmental, technical, social and economic). Nevertheless, the MCDA provided a scientific base for the decision-makers when considering all effects of the different proposed solutions.

Based on Chapter 8, the water supply network would perform optimally in the future under the moderate growth scenario (3% annually local and 4% tourism). However, for the fast and the very fast growth scenarios (5 - 7% local and 7 - 9% tourism, respectively) the water supply system will perform poorly, since the network will need an extensive renovation and expansion. This chapter also showed that spillage of water from household storage tanks accounts for a significant amount of water that is wasted by the population (ranging from 4 to 35% from the total supply, considering 16 scenarios). The results of this chapter also suggest that storage facilities on household premises are not necessarily contributing to the provision of water, but are instead complicating the situation as individual storage tanks. This storage can also create inequity of water availability.

Based on the EPANET model, the current network is sufficient to meet the demand of the slow and moderate growth scenarios. Nonetheless, in the moderate growth scenario (year 30), the network is expected to experience difficulties, but only at the peak hour of 06:00. For this reason, in Chapter 8 two solutions were proposed: (1) a partial renovation of the network, avoiding individual storage tanks (2) maintaining the storage facilities – because of their high acceptance – making use of them at peak hours only.

9.3 Key challenges and future recommendations

Santa Cruz Island is obviously experiencing a water supply and demand strain. Because of rapid population growth, the municipality has not been able to provide a proper water supply. The lack of fresh water is one of the main issues, and the municipal supply is brackish water extracted from crevices. As has been demonstrated in this thesis, the water demand (domestic and from hotels) is quite high considering the lack of quality water volume. The excessive wastage and leakage found within premises partially contributes to only having an intermittent water supply. Strict control of overflow of tanks, as well as an increase in the water tariff is needed. With only these two measures, the amount of available water will increase, as well as the supply hours to most likely a continuous system.

Considering the absence of data, the amount of water that can be sustainably extracted from the basal aquifer is questionable. Because there are no solid studies on variations of salinity levels since the operation of the supply system (2010) from the municipal crevice 'La Camiseta', it is not possible to affirm that there would not be disequilibrium in the recharge of the aquifer. There is evidence from other small islands (e.g Cyprus, Malta, Canary Islands, etc.) that shows negative effects of the overexploitation of groundwater, especially in places where marine intrusion and precipitation need a determined balance in the aquifer.

The current municipal plan involves desalination of brackish groundwater from the crevice 'La Camiseta'. However, no relevant studies regarding the change of Total Dissolved Solids over time have been made. Therefore, the variation of salinity levels may become a serious problem for the desalination plant, since water salinity may increase over the next 30 years. If this is the case, the proposed desalination plant will become useless. Furthermore, water resources may deteriorate to a level where they would not even be of irrigation/agriculture quality. Consequently, the current uses of brackish municipal water would decrease, causing a negative impact on water appliances that are not made for saline water, and would create additional financial costs.

As a result, it is more recommendable to install a seawater desalination plant, even with potential rises in the cost of water and environmental threats. It should be noted that it is widely suggested that desalination plants have negative physicochemical and ecological impacts on marine environments, from hypersaline discharges (brine) causing plumes with elevated salinities, which might extend over tens of meters, depending on the salinity of the brine, mixing depth of the sea, etc. There is also concern regarding the use and release of anti-foulants and anti-scalants to maintain plant infrastructure. This means that the discharge site selection is key to reducing the ecological impacts. Furthermore, areas of biological importance should have a complete environmental study of the desalination disposal area in order to minimize potential harmful impacts to marine ecosystems. Energy use is also highly relevant, especially for this case study area. In order to power a seawater desalination plant, more electricity will need to be generated. Current power supplies are provided by a thermal plant that imports fuel from the mainland. This scenario can increase the electricity costs because of fuel importation, as well as CO_2 emissions. With these views of potential environmental impacts of desalination on this island, a wind or solar powered plant could be a more appealing and sustainable option.

It is highly recommended that the islands limit the number of tourists in the future. As shown in this research, a moderate growth rate of the population and tourism is the maximum sustainable rate. With moderate growth, the municipality would not necessarily need to install a desalination plant, but could implement more sustainable and more cost effective options. However, if tourism growth rates do increase, the installation of a seawater desalination plant is inevitable to generate enough quality water resources. Even so, this option could be delayed with the proposed intervention strategies. If tourism and local population increase at very high rates, the water supply network would encounter severe problems to supply sufficient potable water.

The municipality should monitor private water extraction from crevices, as well as the water demand rates, in order to develop realistic management alternatives. Due to the ecological value and uniqueness of the islands, the same regulations of the national mainland cannot be applied. In addition, migration from the mainland should also be controlled, since it places enormous stress on the existing water supply system and further drives the need for a seawater desalination system on the island. Migration policies should be strengthened in order to avoid population increases from the mainland, as has occurred in past decades. This research shows that the island cannot manage water quantity and quality in the fast or very fast scenarios without a seawater desalination plant. The sustainable strategies proposed in this thesis will suffice only in the slow scenario. In the case of the recommended moderate growth scenario, the sustainable proposed strategies would needed to be combined with other measures in order to mitigate the lack of water supply. This also raises the concern that if humans keep exerting rising environmental pressure on the island, the ecosystem could deteriorate and the main attraction of these islands would be lost along with the economic benefits of tourism.

References

Ajbar, A. and Ali, E. M. (2015). "Prediction of municipal water production in touristic Mecca City in Saudi Arabia using neural networks." Journal of King Saud University-Engineering Sciences **27**(1): 83-91.

Al-Karaghouli, A. and Kazmerski, L. L. (2013). "Energy consumption and water production cost of conventional and renewable-energy-powered desalination processes." Renewable and Sustainable Energy Reviews **24**: 343-356.

Amador, E., Bliemsrieder, M., Cayot, L., Cifuentes, M., Cruz, E., Cruz, F., Rodríguez, J. and Ayora, P. (1996). Plan de Manejo del Parque Nacional Galápagos Servicio Parque Nacional Galápagos, Instituto Ecuatoriano Forestal y de Áreas Naturales y Vida Silvestre.

Ameyaw, E. E., Memon, F. A. and Bicik, J. (2013). "Improving equity in intermittent water supply systems." Journal of Water Supply: Research and Technology-Aqua **62**(8): 552-562.

Andey, S. P. and Kelkar, P. S. (2009). "Influence of intermittent and continuous modes of water supply on domestic water consumption." Water Resources Management **23**(12): 2555-2566.

Aquaprojekt and Hidroekspert (2001). Gospodarsko-tehnicka analiza : Vodoopskrbni sustav otoka Visa - Program prioritetnih radova. Split, Croatia, Croatian Waters

Arbués, F., Garcıa-Valiñas, M. Á. and Martınez-Espiñeira, R. (2003). "Estimation of residential water demand: a state-of-the-art review." The Journal of Socio-Economics **32**(1): 81-102.

Axiak, V., Gauci, V., Mallia, A., Mallia, E., Schembri, P. J., Vella, A. J. and Vella, L. (2002). "State of the Environment Report for Malta 2002." Ministry for Home Affairs and the Environment, Malta.

Beal, C., Stewart, R., Huang, T. and Rey, E. (2011). "SEQ residential end use study." Journal of the Australian Water Association **38**(1): 80-84.

Behzadian, K. and Kapelan, Z. (2015). "Advantages of integrated and sustainability based assessment for metabolism based strategic planning of urban water systems." Science of the total environment **527**: 220-231.

Behzadian, K. and Kapelan, Z. (2015). "Modelling metabolism based performance of an urban water system using WaterMet 2." Resources, Conservation and Recycling **99**: 84-99.

Behzadian, K., Kapelan, Z., Govindarajan, V., Brattebø, H. and Sægrov, S. (2014). "WaterMet2: a tool for integrated analysis of sustainability-based performance of urban water systems."

Behzadian, K., Kapelan, Z., Govindarajan, V., Brattebo, H., Saegrov, S., Rozos, E. and Makropoulos, C. (2014). Quantitative UWS Performance Model: WaterMet^2, TRUST.

Behzadian, K., Kapelan, Z., Venkatesh, G., Brattebo, H., Saegrov, S., Makropoulos, C., Ugarelli, R., Milina, J. and Hem, L. (2013). Urban Water System Metabolism Assessment using WaterMet2 Model. 12th International Conference in Computing and Control for the Water Industry.

Billings, R. B. and Jones, C. V. (2008). Forecasting urban water demand, American Water Works Association.

Boehler, M., Joss, A., Buetzer, S., Holzapfel, M., Mooser, H. and Siegrist, H. (2007). "Treatment of toilet wastewater for reuse in a membrane bioreactor." Water Science and Technology 56(5): 63-70.

Bonacci, O., Ljubenkov, I. and Knezić, S. (2012). "The water on a small karst island: the island of Korčula (Croatia) as an example." Environmental Earth Sciences 66(5): 1345-1357.

Bougadis, J., Adamowski, K. and Diduch, R. (2005). "Short-term municipal water demand forecasting." Hydrological Processes 19(1): 137-148.

Bramwell, B. (2003). "Maltese responses to tourism." Annals of Tourism Research 30(3): 581-605.

Brick, T., Primrose, B., Chandrasekhar, R., Roy, S., Muliyil, J. and Kang, G. (2004). "Water contamination in urban south India: Household storage practices and their implications for water safety and enteric infections." International journal of hygiene and environmental health 207(5): 473-480.

Briguglio, L. (1995). "Small island developing states and their economic vulnerabilities." World development 23(9): 1615-1632.

Briguglio, L. (2008). "Sustainable tourism on small island jurisdictions with special reference to Malta." ARA Journal of Tourism Research 1(1): 29-39.

Bruce, B., Allen, D., Chaves, H., Grant, G., Essink, G. O., Kooi, H., White, I., Gurdak, J., Famiglietti, J. and Martin-Bordes, J. L. (2008). Groundwater Resources Assessment under the Pressures of Humanity and Climate Changes, UNESCO-IHP.

Castillo-Martinez, A., Gutierrez-Escolar, A., Gutierrez-Martinez, J.-M., Gomez-Pulido, J. and Garcia-Lopez, E. (2014). "Water Label to Improve Water Billing in Spanish Households." Water 6(5): 1467-1481.

CDF (2016). "Conservation and Management." Retrieved 11-2016, 2016, from http://www.darwinfoundation.org/en/science-research/conservation-management/.

CGG, C. d. G. d. G.-. (2010). Plan de Gestion Integral del Recurso Hídrico para la Provincia de Galápagos Santa Cruz-Galapagos-Ecuador, Consego de Gobierno de Galápagos.

Charara, N., Cashman, A., Bonnell, R. and Gehr, R. (2010). "Water use efficiency in the hotel sector of Barbados." Journal of Sustainable Tourism 19(2): 231-245.

Cheung, P., Van Zyl, J. and Reis, L. (2005). "Extension of EPANET for pressure driven demand modeling in water distribution system." Computing and Control for the Water Industry 1: 311-316.

Chowdhury, M., Ahmed, M. and Gaffar, M. (2002). "Management of nonrevenue water in four cities of Bangladesh." American Water Works Association. Journal 94(8): 64.

Committee, W. A. D. (2011). Seawater desalination costs White Paper.

d'Ozouville, N. and Merlen, G. (2007). "Agua Dulce o la supervivencia en Galápagos." Galápagos: Migraciones, economía, cultura, conflictos y acuerdos. Biblioteca de Ciencias Sociales 57: 297-313.

d'Ozouville, N., Deffontaines, B., Benveniste, J., Wegmüller, U., Violette, S. and De Marsily, G. (2008b). "DEM generation using ASAR (ENVISAT) for addressing the lack of freshwater ecosystems management, Santa Cruz Island, Galapagos." Remote Sensing of Environment 112(11): 4131-4147.

d'Ozouville, N., Auken, E., Sorensen, K., Violette, S., De Marsily, G., Deffontaines, B. and Merlen, G. (2008a). "Extensive perched aquifer and structural implications revealed by 3D resistivity mapping in a Galapagos volcano." Earth and Planetary Science Letters 269(3): 518-522.

Davis, W. (2003). "Water demand forecast methodology for California water planning areas-work plan and model review." Planning and Management Consultants, Ltd., Carbondale, IL.

De Marchis, M., Fontanazza, C., Freni, G., La Loggia, G., Napoli, E. and Notaro, V. (2010). "A model of the filling process of an intermittent distribution network." Urban Water Journal 7(6): 321-333.

Deng, S.-M. and Burnett, J. (2002). "Water use in hotels in Hong Kong." International Journal of Hospitality Management 21(1): 57-66.

DeVault, G. (2014, 2014). "Surveys Research - Confidence Intervals: Good Survey Research Design Seeks to Reduce Sampling Error." Retrieved 15/07/2014, 2014.

Dhakal, N., Salinas Rodriguez, S., Schippers, J. and Kennedy, M. (2014). "Perspectives and challenges for desalination in developing countries." IDA Journal of Desalination and Water Reuse 6(1): 10-14.

Dinar, A. (1998). "Water policy reforms: information needs and implementation obstacles." Water Policy 1(4): 367-382.

Direccion del Parque Nacional Galapagos , D. (2014). "Statistics of Visitors to Galapagos." Retrieved 01/02/2014, 2014.

Donkor, E. A., Mazzuchi, T. A., Soyer, R. and Alan Roberson, J. (2012). "Urban water demand forecasting: review of methods and models." Journal of Water Resources Planning and Management 140(2): 146-159.

dos Santos, C. C. and Pereira Filho, A. J. (2014). "Water Demand Forecasting Model for the Metropolitan Area of São Paulo, Brazil." Water Resources Management 28(13): 4401-4414.

Ekwue, E. I. (2010). "Management of Water Demand in the Caribbean Region: Current Practices and Future Needs." Management 32(1&2): 28-35.

EPA (2012, 12/19/2012). "Green Building." Retrieved 20/09/2014, 2014, from http://www.epa.gov/greenhomes/bathroom.htm.

Epler, B. (2007). "Tourism, the economy, population growth, and conservation in Galapagos." Charles Darwin Foundation.

Essex, S., Kent, M. and Newnham, R. (2004). "Tourism development in Mallorca: is water supply a constraint?" Journal of Sustainable Tourism 12(1): 4-28.

Fernández Sánchez, M. A. (2013). Análisis de Factibilidad para la Creación de un resort denominado Eco-Scenarium en el km. 4 ½ vía Baltra en la Isla Santa Cruz, Galápagos.

Fidar, A., Memon, F. and Butler, D. (2010). "Environmental implications of water efficient microcomponents in residential buildings." Science of the total environment **408**(23): 5828-5835.

Fletcher, H., Mackley, T. and Judd, S. (2007). "The cost of a package plant membrane bioreactor." Water research **41**(12): 2627-2635.

Fountoulakis, M., Markakis, N., Petousi, I. and Manios, T. (2016). "Single house on-site grey water treatment using a submerged membrane bioreactor for toilet flushing." Science of the total environment **551**: 706-711.

Franco, L. A. and Montibeller, G. (2010). "Problem structuring for multicriteria decision analysis interventions." Wiley Encyclopedia of Operations Research and Management Science.

GADMSC (2012a). Atlás Geográfico del Cantón Santa Cruz. Santa Cruz- Galápagos, Secretaria Técnica de Planificación y Desarrollo Sustentable del Gobierno Autonomo Municipal Descentralizado de Santa Cruz. **1**: 50.

GADMSC (2012b). Plan de Desarrollo y Ordenamiento Territorial (2012-2017). Santa Cruz-Galápagos, Fundación Santiago de Guayaquil, Universidad Católica de Santiago de Guayaqui, Conservación Internacional, AME Ecuador. **1**: 470.

Germanopoulos, G. (1985). "A technical note on the inclusion of pressure dependent demand and leakage terms in water supply network models." Civil Engineering Systems **2**(3): 171-179.

Ghaffour, N., Missimer, T. M. and Amy, G. L. (2013). "Technical review and evaluation of the economics of water desalination: current and future challenges for better water supply sustainability." Desalination **309**: 197-207.

Ghina, F. (2003). "Sustainable development in small island developing states." Environment, Development and Sustainability **5**(1-2): 139-165.

Giampietro, M., Ramos Martín, J., Sorman, A. and Martínez, C. (2012). Impacto de las actividades humanas sobre los ecosistemas locales: Análisis Integrado Multi-Escala del Metabolismo de la Sociedad y del Ecosistema (MuSIASEM). Santa Cruz-Galápagos, Ministerio de Turismo: 80.

Gikas, P. and Tchobanoglous, G. (2009). "Sustainable use of water in the Aegean Islands." Journal of Environmental Management **90**(8): 2601-2611.

Gingerich, S. B. and Oki, D. S. (2000). Ground water in Hawaii, Geological Survey (US).

Gnirss, R., Luedicke, C., Vocks, M. and Lesjean, B. (2008). "Design criteria for semi-central sanitation with low pressure network and membrane bioreactor–the ENREM project." Water Science and Technology **57**(3): 403-410.

González, J. A., Montes, C., Rodríguez, J. and Tapia, W. (2008). "Rethinking the Galapagos Islands as a complex social-ecological system: implications for conservation and management." Ecology and Society **13**(2): 13.

Gössling, S. (2001). "The consequences of tourism for sustainable water use on a tropical island: Zanzibar, Tanzania." Journal of Environmental Management **61**(2): 179-191.

Gössling, S., Hall, C. M. and Scott, D. (2015). Tourism and water, Channel View Publications.

Gössling, S., Peeters, P., Hall, C. M., Ceron, J.-P., Dubois, G., Lehmann, L. V. and Scott, D. (2012). "Tourism and water use: Supply, demand, and security. An international review." Tourism Management **33**(1): 1-15.

Guilabert Antón, L. (2012). La gestión del abastecimiento de agua a las ciudades: El caso de Tenerife. Gestión Sostenible y Tecnologías del Agua. Alicante, Universidad de Alicante. **Master Oficial:** 71.

Guitouni, A. and Martel, J.-M. (1998). "Tentative guidelines to help choosing an appropriate MCDA method." European Journal of Operational Research **109**(2): 501-521.

Gupta, R. and Bhave, P. R. (1996). "Comparison of methods for predicting deficient-network performance." Journal of Water Resources Planning and Management **122**(3): 214-217.

Guyot-Tephiane, J., Daniel Orellana, Christopher Grenier (2012). Informe científico de la campaña de encuesta "Percepciones, Usos y Manejo del agua en Galápagos". Santa Cruz-Galápagos, Fundación Charles Darwin & Universidad de Nantes. **1.**

Hauber-Davidson, G. and Shortt, J. (2011). "Energy Consumption of Domestic Rainwater Tanks." Water Journal **38**(3): 1-5.

Hayuti, M. and Burrows, R. (2004). "Sequential solution seeking dda based hda (sss-dda/hda) approach." Proc. of Decision Support in the Water Industry under Conditions of Uncertainty-ACTUI 2004.

Herwijnen, M. v. and Janssen, R. (2004). "Software support for multi-criteria decision making." Sustainable Management of Water Resources: an integrated approach.

Hof, A. and Schmitt, T. (2011). "Urban and tourist land use patterns and water consumption: Evidence from Mallorca, Balearic Islands." Land Use Policy 28(4): 792-804.

Hophmayer-Tockich, S. and Kadiman, T. (2006). Water Management on Islands–Common Issues and Possible Actions: A concept paper in preparation to the international workshop:'Capacity Building in Water Management for Sustainable Tourism on Islands'. Twente-The Netherlands, Department of Governance and Technology for Sustatainability-University of Twente.

INAMHI (2014). "Instituto Nacional de Meteorología e Hidrología." Retrieved 05/12/2014, 2014, from http://www.serviciometeorologico.gob.ec/.

INEC (2010). Censo de Población y vivienda del Ecuador 2010. Ecuador, Instituto Nacional de Estadísticas y Censos.

Jansen, A. and Schulz, C. e. (2006). "Water demand and the urban poor: A study of the factors influencing water consumption among housholds in Cape town, South Africa." South African Journal of Economics 74(3): 593-609.

Janssen, R. and van Herwijnen, M. (2011). "Definite 3.1." Vrije Universiteit Amsterdam, IVM.

Janssen, R., Van Herwijnen, M. and Beinat, E. (2001). "DEFINITE for Windows." A system to support decisions on a finite set of alternatives (Software package and user manual) Institute for Environmental Studies (IVM), Vrije Universiteit, Amsterdam.

Kechagias, E. and Katsifarakis, K. L. (2004). "Planning Water Resources Management in Small Islands. The Case of Kalymnos, Greece." Water, Air and Soil Pollution: Focus 4(4-5): 279-288.

Khaka, E. (1998). "Small islands, big problems." Our Planet 9: 25-26.

Kiker, G. A., Bridges, T. S., Varghese, A., Seager, T. P. and Linkov, I. (2005). "Application of multicriteria decision analysis in environmental decision making." Integrated environmental assessment and management 1(2): 95-108.

Kondili, E., Kaldellis, J. K. and Papapostolou, C. (2010). "A novel systemic approach to water resources optimisation in areas with limited water resources." Desalination 250(1): 297-301.

Konstantopoulou, F., Liu, S., Papageorgiou, L. and Gikas, P. (2011). "Water Resources Management for Paros Island, Greece." International Journal of Sustainable Water and Environmental Systems 2(1): 1-6.

Last, E. M. (2010). City Water Balance A New Scoping Tool For Integrated Urban Water Management Options. Birmingham, UK, Unpublished., University of Birmingham. **Doctor of Philosophy**.

Lattemann, S., Kennedy, M. D., Schippers, J. C. and Amy, G. (2010). "Global desalination situation." Sustainability Science and Engineering 2: 7-39.

Linkov, I., Satterstrom, F., Kiker, G., Batchelor, C., Bridges, T. and Ferguson, E. (2006). "From comparative risk assessment to multi-criteria decision analysis and adaptive management: Recent developments and applications." Environment International 32(8): 1072-1093.

Liu, J. (2011). Investigación de la Calidad Bacteriológica del Agua y de las Enfermedades Relacionadas al Agua en la Isla Santa Cruz - Galápagos. Santa Cruz- Galápagos, Fundación Charles Darwin, Comisión Fullbright. **1**.

Liu, J. and d'Ozouville, N. (2013). "Water contamination in Puerto Ayora: Applied interdisciplinary research using Escherichia coli as an indicator bacteria." GALAPAGOS REPORT 2011-2012: 76.

Liu, S., Papageorgiou, L. G. and Gikas, P. (2012). "Integrated management of non-conventional water resources in anhydrous islands." Water Resources Management 26(2): 359-375.

López, J. and Rueda, D. (2009). "Water quality monitoring system in Santa Cruz, San Cristóbal, and Isabela." Galapagos Report 2010-2011.

Makropoulos, C. K., Natsis, K., Liu, S., Mittas, K. and Butler, D. (2008). "Decision support for sustainable option selection in integrated urban water management." Environmental Modelling & Software 23(12): 1448-1460.

Mangion, E. (2013). Tourism impact on water consumption in Malta The Smeed Report and Road Pricing: The Case of Valletta, Malta: 61.

Marques, R. C., Simões, P. and Berg, S. (2013). "Water sector regulation in small island developing states: an application to Cape Verde." Water Policy 15(1): 153-169.

Martinez, F., Signes, M., Savall, R., Andrés, M., Ponz, R. and Conejos, P. (1999). "Construction and use of dynamic simulation model for the valencia metropolitan water supply and distribution network." Water Industry Systems: Modeling and Optimization Applications, In: Savic, DA and Walters, GA (eds.) **1**: 155-174.

Mena, C., Walsh, S., Pizzitutti, F., Reck, G., Rindfuss, R., D., O., Toral-Granda, V., Valle, C., Quiroga, D., García, J., Vasconez, L., Guevara, A., Sanchez, M., Frizelle, B. and Tippett, R. (2013). Determination of social, environmental and economical relations which allow the development based on different processes of modeling, potential scenarios of sustainability of the socio-ecological system of the Galapagos Islands with emphasis on the dynamic of the flux of tourist visitors. Galapagos, Ministry of Environment & Galapagos National Park. **1**.

Mitchell, G. and Diaper, C. (2010). UVQ User Manual, CSIRO Urban Water System and Technologies.

Mitchell, V., Duncan, H., Inma, R. M., Stewart, J., Vieritz, A., Holt, P., Grant, A., Fletcher, T., Coleman, J. and Maheepala, S. (2007). "State of the art review of integrated urban water models." Novatech Lyon, France: 1-8.

Mutikanga, H. E., Sharma, S. K. and Vairavamoorthy, K. (2011). "Multi-criteria decision analysis: a strategic planning tool for water loss management." Water Resources Management **25**(14): 3947-3969.

Nurse, L. A., Sem, G., Hay, J., Suarez, A., Wong, P. P., Briguglio, L. and Ragoonaden, S. (2001). "Small island states." Climate change: 843-875.

OECD (2009). Managing water for all: an OECD perspective on pricing and financing, Organisation for Economic Co-operation and Development Publisher.

Ortiz, J. (2006). Aquifers in Santa Cruz Island-Galápagos. Geography Department. London, King's College London. **Master in Science**.

Oswald, W. E., Lescano, A. G., Bern, C., Calderon, M. M., Cabrera, L. and Gilman, R. H. (2007). "Fecal contamination of drinking water within peri-urban households, Lima, Peru." The American journal of tropical medicine and hygiene **77**(4): 699-704.

Polatidis, H., Haralambopoulos, D. A., Munda, G. and Vreeker, R. (2006). "Selecting an appropriate multi-criteria decision analysis technique for renewable energy planning." Energy Sources, Part B **1**(2): 181-193.

Pryet, A. (2011). Hydrogeology of volcanic islands: a case-study in the Galapagos Archipelago (Ecuador), Paris 6.

Redfern, P. a. (2003). Estudio de provisión de agua y tratamiento de aguas residuales de Santa Cruz. Santa Cruz-Galápagos, Proctor and Redfern International Limited.

Retamal, M., Turner, A. and White, S. (2009). "Energy implications of household rainwater systems." Australian Water Association 36(8): 70-75.

Reyes, M., Trifunovic, N., Saroj, S. and Kennedy, M. (2015). Water supply and demand in Santa Cruz Island/Galápagos Archipelago XVth International Water Technology Conference. Sharm El Sheikh-Egypt, International Water Techonology Association.

Reyes, M., Trifunovic, N., d'Ozouville, N., Sharma, S. and Kennedy, M. (2017). "Quantification of urban water demand in the Island of Santa Cruz (Galápagos Archipelago)." Desalination and Water Treatment 64: 1-11.

Reyes, M. F., Trifunović, N., Sharma, S. and Kennedy, M. (2016). "Data assessment for water demand and supply balance on the island of Santa Cruz (Galápagos Islands)." Desalination and Water Treatment 57(45): 21335-21349.

Richardson, R. (2017). "Island Ecosystems." Retrieved 01-2017, 2017, from https://www.iucn.org/commissions/commission-ecosystem-management/our-work/cems-specialist-groups/island-ecosystems.

Rogers, P., De Silva, R. and Bhatia, R. (2002). "Water is an economic good: How to use prices to promote equity, efficiency, and sustainability." Water Policy 4(1): 1-17.

Rossman, L. A. (2000). EPANET 2: users manual.

Rozos, E., Makropoulos, C. and Butler, D. (2009). "Design robustness of local water-recycling schemes." Journal of Water Resources Planning and Management 136(5): 531-538.

Salaguste-Anarna, M. (2009). Development of a multi-criteria decision support system for seawater desalination plants, Unesco-IHE.

Salminen, P., Hokkanen, J. and Lahdelma, R. (1998). "Comparing multicriteria methods in the context of environmental problems." European Journal of Operational Research 104(3): 485-496.

Sarango, D. (2013). Interview water supply sytem. M. F. Reyes.

Seetharam, K. and Bridges, G. (2005). "Helping India Achieve 24x7 Water Supply Service by 2010." Roundtable Discussion on Private Sector Participation in Water Supply in India, June: 15-16.

Sharma, S. K. and Vairavamoorthy, K. (2009). "Urban water demand management: prospects and challenges for the developing countries." Water and Environment Journal 23(3): 210-218.

Simpson, L. (1996). "Do decision makers know what they prefer?: MAVT and ELECTRE II." Journal of the Operational Research Society 47(7): 919-929.

Singh, O. and Turkiya, S. (2013). "A survey of household domestic water consumption patterns in rural semi-arid village, India." GeoJournal 78(5): 777-790.

Soares, A., Reis, L. and Carrijo, I. (2003). "Head-driven simulation model (HDSM) for water distribution system calibration." Advances in Water Supply Manament, Maksimovic, C.; Butler, D.; Memon, FA (eds.), Swets and Zeillinger, Lisse 1: 197-207.

Soyer, R. and Roberson, J. A. "Urban Water Demand Forecasting: A Review of Methods and Models."

Tam, V. W., Tam, L. and Zeng, S. (2010). "Cost effectiveness and tradeoff on the use of rainwater tank: An empirical study in Australian residential decision-making." Resources, Conservation and Recycling 54(3): 178-186.

Tanyimboh, T. (2004). "Availability of water in distribution systems."

Tanyimboh, T., Tabesh, M. and Burrows, R. (2001). "Appraisal of source head methods for calculating reliability of water distribution networks." Journal of Water Resources Planning and Management 127(4): 206-213.

Trifunovic, N. (2006). Introduction to Urban Water Distribution: Unesco-IHE Lecture Note Series, CRC Press.

Trueman, M. and d'Ozouville, N. (2010). "Characterizing the Galapagos terrestrial climate in the face of global climate change." Galapagos Research 67: 26-37.

TRUST (2014). "Trust." Retrieved 2 October 2014, 2014, from http://www.trust-i.net/.

Tsiourtis, N. X. (2002). "Water charge, the Cyprus experience." Water Valuation and Cost Recovery Mechanisms in the Developing Countries of the Mediterranean Region. CIHEAM-IAMB, Bari: 91-104.

UNESCO (2009). The United Nations World Water Development Report 3: Water in a Changing World. London, United Nations Educational, Scientific and Cultural Organization (UNESCO).

UNICEF, W. (2000). Global water supply and sanitation assessment 2000 Report, World Health Organization.

Vairavamoorthy and Elango (2002). "Guidelines for the design and control of intermittent water distribution systems." Waterlines 21(1): 19-21.

Vairavamoorthy, K., Gorantiwar, S. D. and Mohan, S. (2007). "Intermittent water supply under water scarcity situations." Water international 32(1): 121-132.

Vairavamoorthy, K., Gorantiwar, S. D. and Pathirana, A. (2008). "Managing urban water supplies in developing countries–Climate change and water scarcity scenarios." Physics and Chemistry of the Earth, Parts A/B/C 33(5): 330-339.

Verrecht, B., Maere, T., Benedetti, L., Nopens, I. and Judd, S. (2010). "Model-based energy optimisation of a small-scale decentralised membrane bioreactor for urban reuse." Water research 44(14): 4047-4056.

Violette, S., d'Ozouville, N., Pryet, A., Deffontaines, B., Fortin, J. and Adelinet, M. (2014). "Hydrogeology of the Galapagos Archipelago: an integrated and comparative approach between islands." The Galapagos: A Natural Laboratory for the Earth Sciences 204: 167.

Wang, J.-J., Jing, Y.-Y., Zhang, C.-F. and Zhao, J.-H. (2009). "Review on multi-criteria decision analysis aid in sustainable energy decision-making." Renewable and Sustainable Energy Reviews 13(9): 2263-2278.

Watkins, K. (2006). "Human Development Report 2006-Beyond scarcity: Power, poverty and the global water crisis." UNDP Human Development Reports (2006).

White, I. and Falkland, T. (2010). "Management of freshwater lenses on small Pacific islands." Hydrogeology Journal 18(1): 227-246.

WMI-GIZ, P. (2013). Mejoramiento del sistema municipal de distribución de agua y reducción de las pérdidas informe ejecutivo – misión operativa n°4. Santa Cruz-Galápagos, Water Management International, Gobierno Autonomo Descentralizado Municipal de Santa Cruz.

Wright, J., Gundry, S. and Conroy, R. (2004). "Household drinking water in developing countries: a systematic review of microbiological contamination between source and point-of-use." Tropical Medicine & International Health 9(1): 106-117.

Yatsalo, B., Didenko, V., Gritsyuk, S. and Sullivan, T. (2015). "Decerns: a framework for multi-criteria decision analysis." International Journal of Computational Intelligence Systems 8(3): 467-489.

Uncategorized References

d'Ozouville, N. (2009). Manejo de recursos hídricos: Caso de la cuenca de Pelican Bay. Galapagos Report 2007-2008. Puerto-Ayora, Santa Cruz, Galpagos-Ecuador, Imprenta Monsalve Moreno.

Donta, A. and Lange, M. (2008). Water management on Mediterranean Islands: pressure, recommended policy and management options. Coping with water deficiency, Springer: 11-44.

Ingeduld, P., Svitak, Z., Pradhan, A. and Tarai, A. (2006). Modelling intermittent water supply systems with EPANET. 8th annual water distribution systems analysis symposium, Cincinnati.

Ozger, S. S. and Mays, L. (2003). A semi-pressure-driven approach to reliability assessment of water distribution networks. Proceedings of the Thirtieth Congress.

Pathirana, A. (2010). EPANET2 desktop application for pressure driven demand modeling. Water Distribution Systems Analysis 2010: 65-74.

SENAGUA (2012). Propuesta Participativa de la Política Pública Nacional del Agua del Ecuador. Ecuador, SENAGUA.

Sharma, S. (2014). Urban Water Supply and Demand Management T. N. UNESCO-IHE Institute for water education.

Trifunovic, N. (2008). Introduction to Urban Water Distribution. UNESCO-IHE, The Netherlands, Taylor & Francis/Balkema, Leiden, The Netherlands: 527.

Trifunović, N. and Abu-Madi, M. O. R. (1999). Demand modelling of networks with individual storage. WRPMD'99: Preparing for the 21st Century: 1-10.

List of abbreviations and symbols

ANN-Artificial Neural Networks

B-Benefit

BWRO-Brackish Water Reverse Osmosis

CFU –Colony Forming Unit

C- Cost

CGG-Consejo de Gobierno de Galápagos

DDA-Demand Driven Analysis

DPNG- Dirección del Parque Nacional Galápagos

DPWS- Department of Potable Water and Sewarage

EPA-Environmental Protection Agency

GADMSC-Gobierno Autónomo Descentralizado del Municipio de Santa Cruz

GSC- Galápagos Science Center

GWR- Greywater Recycling

HP-Horse Power

IBY- Increased Block Tariff

INAMHI-Instituto Nacional de Meteorología e Hidrología

INEC-Instituto Ecuatoriana de Estadísticas y Censos

KPI- Key Performance Indicator

KWh- Kilowatt Hour

L- Litters

lpcpd- litters per capita per day

L/s-Litter per second

MAUT- Multi Attribute Utility Theory

MBR- Membrane Brioreactor

MCDA -Multi-Criteria Decision Analysis

mg/L- miligrams per litter

MINTUR- Ministry of Tourism of Ecuador

m^3-cubic meter

mm-milimiters

NGO- Non-Governmental Organizations

NRW - Non-Revenue Water

O&M- Operations and Management

OECD- Organization for Economic and Co-operation and Development

OLS-Ordinary Least Squares

PDA- Pressure Driven Analysis

PDLS - People driven levels of service

PIGRH - Plan Regional de Gestión de Recursos Hídricos

PVC-Polyvinylchloride

PZ 1_ Pilot zone 1

PZ 2- Pilot zone 2

PZ 3- Pilot zone 3

RO-Reverse Osmosis

RWH-Rainwater Harvesting

SENAGUA- Secretaría Nacional del Agua

SIDS - Small Island Developing States

SWRO-Seawater Reverse Osmosis

USD-Dollars from the United States

USFQ-Universidad San Francisco de Quito

UWDM -Urban water demand measures

UWS-Urban Water System

WDM- Water Demand Management

WMI- Water Management International

WTC-Water supply conduits

WTW-water treatment works

WWF- World Wildlife Foundation

WWTW-Wastewater Treatment Works

Annexes

ENTREVISTA PARA CONSUMIDORES DOMESTICOS

Género del encuestado/a: F___ M___   **Barrio donde vive: _____**

Dirección: _____

1) **Edad del encuestado/a:**

   16-25 años ___   b. 26-35 años___ c. 36-45 años___

   d. 46-55 años____   e. Más de 55___

2) *En qué tipo de vivienda habita?*

Suite	
Departamento	
Casa	
Otro (precisar)	

3) *Cuál es el número de habitaciones en la propiedad (incluyendo cocina y sala)?*

   a. 1___   b. 2 ___   c.3___   d.4 ___   e. 5 o más___

4) *Cuántas personas viven en esta propiedad/domicilio?*

   a. 1___   b. 2 ___   c.3___   d.4 ___   e. 5 o más___

5) *Cuántas personas en esta propiedad están empleadas?*

   a. 1___   b. 2 ___   c.3___   d.4 ___   e. 5 o más___

6) *A qué horas salen usualmente las personas empleadas al trabajo?*

   a. De 6h00 a 7h00___   b. De 7h01 a 8h00___   c. De 8h01 a 9h00___

   d. De 9h01 a 10h00___   e. De 10h00 en adelante___   f. No Sabe/No contesta___

7) *A qué horas salen usualmente los otros miembros de la familia a la escuela/colegio/universidad/otros?*

   a. De 6h00 a 7h00____   b. De 7h01 a 8h00___   c. De 8h01 a 9h00 ___

   d. De 9h01 a 10h00 ____   e. De 10h00 en adelante____   f. No Sabe/No contesta___

8) *Cuántas personas regresan usualmente a la casa para el almuerzo?*

   a. 1___   b. 2 ___   c.3___   d.4 ___   e. 5 o más___

9) *A qué hora usualmente regresan las personas a la casa para el almuerzo?*

   a.De 12h00 a 12h30 ___   b. De 12h31 a 13h00___   c. De 13h01 a 13h30___

d. De 13h31 a 14h00___   e. De 14h00 en adelante___   f. No Sabe/No contesta___

10) A qué horas regresan usualmente a la casa las personas empleadas cuando termina la jornada laboral?

a.De 16h00 a 17h00___   b. De 17h01 a 18h00___   c. De 18h01 a 19h00___

d.De 19h01 a 20h00 ___   e. De 20h00 en adelante___   f. No Sabe/No contesta___

11) A qué horas regresan usualmente a la casa los estudiantes/otros cuando terminan la jornada de la escuela/universidad/otros?

    a.   Antes de las 16h00___   b. De 16h00 a 17h00___   c. De 17h01 a 18h00___

    d.   De 18h01 a 19h00___   e. De 19h01 a 20h00___   f. De 20h00 en adelante___

    g.   No Sabe/No contesta___

12) La comida (almuerzo/cena) es preparada en casa?

Si ____ NO____ No Sabe/No Contesta____

- *Si sí, cuantas veces al día?*

a.1___   b.2____   c.3____   d.4 o más___   e. No Sabe/No contesta___

12) Cuántos de los siguientes ítems que utilizan agua tiene dentro de su domicilio?

Item	No. of items
Inodoro	
Lavabo de bano	
Lavabo de cocina	
Ducha	
Tina	
Sistema de riego para jardín	
Lavadora de ropa	
Lavadora de platos	
Otros (Especificar)	

13) Usted está conectado a la red municipal de abastecimiento de agua?

Si ____ NO____ No Sabe/No Contesta____

14) Con qué frecuencia recibe agua municipal (escoger solo una respuesta de a,b,c)?

    a.   Todos los días___

    - Especificar número de horas

i.1 hora___   ii. 2 horas___   iii. 3 horas___   iv. 4 horas___   v.5 horas o más___

    b.   Pasando un día ___

    -Especificar número de horas diarias

i.1 hora___   ii. 2 horas____   iii. 3 horas___   iv. 4 horas___   v.5 horas o más___

    c.   Pasando dos días ___

-Especificar el número de horas

i.1 hora___   ii. 2 horas_____   iii. 3 horas___   iv. 4 horas___   v.5 horas o más___

15) **Para qué tipo de actividades domésticas utiliza el agua provista por el sistema municipal?**

Inodoros	
Ducha	
Lavar platos	
Cocinar	
Beber	
Aseo personal (dientes, cara, etc)	
Lavar ropa	
Otros (especificar)	

16) **Qué tarifa fija paga al municipio por el servicio de agua (Cuál fue el pago de la planilla del mes pasado)?**

a.Doméstica (5.24 USD) ___   b. Comercial (11.22 USD) ___

c.Industrial/residencial (28.50 USD   d. Industrial Hoteles (45,00 USD) ____

e. No Paga___

17) **Si es que cuenta con medidor (Bellavista) cuál fue el pago del mes anterior?**

a.Entre 5 USD y 10 USD___   b. Entre 11 USD y 15 USD___   c. Entre 16 USD y 20 USD___

d.Entre 21 USD y 25 USD___   e. 26 USD o más___   f. No Sabe/No contesta___

18) **Tiene usted algún sistema de almacenamiento/cisterna para el agua municipal?**

Si ____   NO_____   No Sabe/No Contesta___

- Qué tipo de almacenamiento es:

a. Cisterna_____   b. Tanque Elevado_____   c. Cisterna y Tanque Elevado___   d. Otros (Especificar)_____

- Cuál es el volúmen aproximado del tanque/cisterna (en total)?

a.1 m3_____   b. 2 m3___   c. 3 m3___   d.4 m3___   e.5 m3 o más ___   f. No sabe

- Cuantas veces se llena a la semana el total del volumen?

a.1 vez_____   b. 2 veces___   c. 3 veces___   d.4 veces___   e. 5 veces o más ___   f. No sabe

19) **Usted espera a que se llene el tanque/cisterna para cerrar las llaves cuando están siendo llenado/s?**

Si ____   NO_____   No Sabe/No Contesta___

20) **Tiene instalado un flotador en su cisterna de almacenamiento que detiene el flujo del agua cuando la cisterna está llena o que evita que se derrame?**

Si ____   NO_____   No Sabe/No Contesta___

21) **Ha identificado fugas en su domicilio?**

Si ____   NO_____   No Sabe/No Contesta___

-Dónde se encuentran principalmente estas fugas? (Escoger solo una respuesta)

a.Inodoros___   b. Cocina___   c. Ducha___   d. Jardín_____   e. Otros___

22) **Usted compra agua de tanquero?**

Si ____ NO ____ No Sabe/No Contesta ____

-Si si, cuántas veces por semana?
a.1 ___   b. 2 ____   c.3 ____   d.4 ____   e.5 o más ____

23) Para qué tipo de actividades domésticas utiliza el agua de tanquero?

Inodoros	
Ducha	
Lavar platos	
Cocinar	
Beber	
Aseo personal (dientes, cara, etc.)	
Otros (Especificar)	

24) Compra usted agua purificada (desalinizada)?

Si ____ NO ____ No Sabe/No Contesta ____

25) Cuántos envases de agua compra por semana?

Envase	Cantidad
Pomas (5 galones)	
Envase de galón	
Botellon Azul (20 L)	
Al granel	

26) Para qué actividades usa el agua purificada?

Inodoros	
Ducha	
Lavar platos	
Cocinar	
Beber	
Aseo personal (dientes, cara, etc)	
Lavar ropa	
Otros (especificar)	

27) Recolecta su hogar agua de lluvia?

Si ____ NO ____ No Sabe/No Contesta ____

28) Puede estimar la capacidad del tanque para recolectar agua de lluvia?

a.1 m ___   b. 2m3 ___   c. 3m3 ____   d. 4m3 ____   e.5m3 o mas ____ f. No sabe ___

29) Para qué tipo de actividades domésticas utiliza el agua lluvia?

Inodoros	
Ducha	
Lavar platos	
Cocinar	
Beber	
Aseo personal	

(dientes, cara, etc.)		
Otros (Especificar)		

30) *Realiza su hogar algún tratamiento al agua ?*

	Sin tratamiento	Hierve	Usa filtros o algún sistema de desinfección
Agua municipal			
Agua purificada			
Agua de lluvia			
Agua de tanquero			

31) *Cuánta gasta por el recurso agua por mes en USD?*

    a. *Agua Municipal*

      1)No Paga___   2) De 5 USD a 10 USD___   3) De 11 USD a 15 USD___

      4)De 16 USD a 20 USD ___   5) Desde 21 USD en adelante ____

    b. *Agua Purificada (Desalinizada)*

      1)No Paga___   2) De 5 USD a 10 USD___   3) De 11 USD a 15 USD___

      4)De 16 USD a 20 USD ___   5) Desde 21 USD en adelante ____

    c. *Agua Lluvia (Inversión Inicial o mantenimiento)*

      1)No Paga___   2) De 5 USD a 10 USD___   3) De 11 USD a 15 USD___

      4)De 16 USD a 20 USD ___   5) Desde 21 USD en adelante ____

    d. *Agua de Tanquero*

      1)No Paga___   2) De 5 USD a 10 USD___   3) De 11 USD a 15 USD___

      4)De 16 USD a 20 USD ___   5) Desde 21 USD en adelante___

32) *Está dispuesto a pagar más por una mejor calidad de agua del sistema municipal?*

    Si ____ NO____ No Sabe/No Contesta___

    *Porqué si o porqué no?*_____

33) *Está dispuesto a pagar por un mejor servicio de agua municipal (que sea continuo)?*

    Si ____ NO____ No Sabe/No Contesta___

34) *Cuánto está dispuesto a pagar máximo por agua que sea potable y por un servicio continuo por mes mediante el sistema municipal?*

    a.De 10 a 20 USD___   b. De 21 a 30 USD____   c. De 31 a 40 USD____   d. Más de 50 USD___

35) *Tendría algún problema de ser colocado un medidor (pagar exactamente lo que consume)?*

    Si ____ NO____ No Sabe/No Contesta___

36) *Usted varía el consumo de agua de acuerdo a la estación del año (menos en garúa (frío) y más en invierno (calor))?*

    Si ____ NO____ No Sabe/No Contesta___

37) *Usted resusa el agua de alguna forma?*

    Si ____ NO____ No Sabe/No Contesta___

-Si sí, para qué usos/actividades reusa el agua?

38) Usted cree que es importante ahorrar agua de alguna forma?

Si ____ NO____ No Sabe/No Contesta____

39) Usted tiene algún aparato/equipo ahorrador de agua?

Si ____ NO____ No Sabe/No Contesta____

  a. Si no, tiene interés en instalar un equipo ahorrador de agua?

  Si ____ NO____ No Sabe/No Contesta____

40) Qué tipo de sistema de saneamiento tiene usted?

Tanque Séptico	
Descarga directa	
Diluye y luego descarga	
Otro (Especificar)	

41) Con qué frecuencia limpia su tanque séptico?

a.Una vez cada seis meses ____ b. Una vez al año ____ c. Una vez cada dos años__

d.Otro (Especificar) ____ e. Nunca lo he limpiado __

CRITERIA/ INDICATORS	COST/BENEFIT cost- negative correlation, benefits- positive correlation	UNIT (ratio.interval.ordinal.binary)	EFFECTS TABLE					STANDARDIZED EFFECTS TABLE				
			GAL1	GAL 2	GAL3	GAL 4	GAL 5	GAL 1	GAL 2	GAL3	GAL 4	GAL 5
ENVIRONMENTAL												
land use	C	(m2)	0	5000	1350	1350	1350	1	0.5	0.86	0.86	0.86
discharge of the waste water	C	(m3/day)	2378.68	43754.83	3034.72	2487.45	2812.23	0.95	0.12	0.94	0.95	0.94
sea water intrusion	C	ordinal scale	1	4	3	3	2	1	0.46	0.79	0.79	0.96
energy consumption	C	(kWh/m3)	3.16	10.04	5.96	4.65	3.55	0.69	0	0.41	0.54	0.65
chemicals use	C	binary scale	no	yes	no	no	no	1	0	1	1	1
impact on endemic species	C	ordinal scale	1	5	4	2	3	1	0.46	0.71	0.96	0.87
impact to marine/land ecosystem	C	ordinal scale	1	5	4	2	3	1	0.46	0.71	0.96	0.87
TECHNICAL												
hours of service or improvement on hours of service	B	binary scale	no	yes	no	no	no	0	1	0	0	0
coverage of demand with supply	B	(% of demand)	45%	100%	56%	50%	59%	0.21	1	0.37	0.29	0.41
water losses	C	(% of the delivered)	21%	15%	9%	25%	11%	0.37	0.68	1	0.16	0.89
robustness of the WS system	B	ordinal scale	5	1	3	2	3	0.46	1	0.79	0.96	0.79
O&M of the WS system	B	ordinal scale	1	4	2	2	3	1	0.46	0.92	0.92	0.71
share of alternative water sources in overall water balance (RW harvesting, GWR)	B	(%)	0%	0%	35%	31%	24%	0	0	0.7	0.62	0.48
Compatibility	B	0/++	+/++	0/+	+	+	0/+	0.75	0.25	0.5	0.5	0.25

CRITERIA/ INDICATORS	COST/BENEFIT cost-negative correlation, benefits-positive correlation	UNIT (ratio, interval, ordinal, binary)	EFFECTS TABLE					STANDARDIZED EFFECTS TABLE				
			GAL1	GAL 2	GAL3	GAL 4	GAL 5	GAL 1	GAL 2	GAL3	GAL 4	GAL 5
ECONOMICAL												
Capital cost	C	(mln €)	8.17	64.88	13.52	7.79	14.74	0.87	0	0.79	0.88	0.77
O&M cost	C	(mln €/year)	0.64	2.00	2.37	1.00	1.26	0.73	0.16	0	0.58	0.47
NRW income generation	B	(€/year)	63,973.93	312,412.16	299,678.89	35,415.01	217,194.67	0.2	1	0.96	0.11	0.7
WDM income generation	B	0/++	+	+	0/+	0/+	+/++	0.5	0.5	0.25	0.25	0.75
Employment creation	B	0/++	0/+	+	0/+	+	+	0.25	0.5	0.25	0.5	0.5
Increase in water tariffs	C	-/0	-/0	-	-	-/0	+	0.75	0	0.5	0.75	0.5
Increase of tourist capacity	B	number	8,425	9,628	8,700	9,628	9,628	0.14	0.3	0.18	0.3	0.3
SOCIAL												
Social acceptability	B	0/+++	0	+	+++	+	++	0	0.33	1	0.33	0.67
Willingness to pay / contribute	B	0/++	+	0/+	+/++	+	+	0.5	0.25	0.75	0.5	0.5
Transparency of project implementation process	B	0/++	+/++	0/+	+	+	+/++	0.75	0.25	0.5	0.5	0.75
water quality improvement	B	0/++	0	++	0/+	0/+	0/+	0	1	0.12	0.12	0.12
Infection risk and annual water related diseases rates	C	-/0	-	0	-/-	-/-	-	0	1	0.12	0.12	0.3
Compatibility to the legislative system	B	binary scale	yes	yes	yes	yes	yes	1	1	1	1	1

Date:

Name:

Place of work and position:

QUESTIONNAIR FOR CRITERIA WEIGHTING OF INTERVENTION STRATEGIES DEVELOPED FOR OPTIMIZING WATER SUPPLY AND DEMAND

Four main criteria have been developed with the purpose of conducting a multi criteria decision analysis. This analysis will be used to evaluate a set of intervention strategies previously developed. The aim is to find the most sustainable solution to solve the current water supply deficits, cover current and future water demand, and find a sustainable balance between supply and demand. The Intervention strategies will be evaluated against four main sets of criteria. These criteria concern the following: **environment, technical aspects of the water supply system, economical aspects and impacts, and social aspects.**

Please rate the importance of the provided criteria based on your personal knowledge, background and experience, with to the following marks:

5 – very important : 4 – important : 3 – neutral : 2 – less important : 1 – not important

It is allowable to assign the same mark to more than one criteria with the same relevance

1. **GENERAL CRITERIA RATING**
   Which of these general criteria do you find more important during the planning and decision making process for an optimal water supply system? Please rate the importance of each criteria considered in this research analysis.

   Environmental (*minimize impact on environment*)   (   )

   Technical (*maximize reliability and accessibility of water supply*)   (   )

   Economical (minimize costs, while maximizing population growth)   (   )

   Social (*maximize public acceptability and compatibility, improve public health*) (   )

2. **ENVIRONMENTAL CRITERIA**
   Please rate the importance of the following sub-criterion in order to minimize impact on the environment:

   Impact to natural resources (*waste disposal, land use*

   discharge of the waste water, sea water intrusion   )   (   )

   Usage of energy and chemicals in infrastructure

   (ratio of fossil fuels against alternative sources, chemical usage)   (   )

   Impact on flora and fauna

   (impact on endemic species, marine ecosystem and land ecosystem)   (   )

3. **TECHNICAL CRITERIA**

Please rate the importance of the following technical criteria in order to provide a water supply system with maximum reliability and accessibility.

Water supply and demand

(improvement on hours of service, coverage of demand with supply, water losses (NRW)) ( )

Infrastructure, operation and maintenance

(percentage of connected consumers, risk of failure of the WS system, O&M of the WS system, share of alternative water sources in overall water balance, compatibility with existing system) ( )

4. **ECONOMICAL ASPECTS**

   Please rate the importance of the following economic criteria in order to minimize the costs, and maximize the benefits and economical growth potential.

   Direct and indirect costs related to the WS system
   (construction, O&M, less income due to the tourism reduction) ( )

   Direct and indirect benefits related to improvement of WS system
   (Income generation due to NRW reduction, and WDM strategy, and improved O&M

   ( )

   Impact to local economy
   (Employment creation, increase in water tariffs, increase of tourist capacity) ( )

5. **SOCIAL ASPECT**

   Please rate the importance of the following criteria in order to maximize social and legal acceptability, and project implementation transparenc.

   Public perception *(social acceptability, willingness to pay and transparency of the project on the implementation process)* ( )

   Public health *(water quality and annual infections risk)* ( )

   Legal and political system *(influence of local politics on project implementation and compatibility with the legislative system)* ( )

6. **Please provide any comment or/and suggestion you find to be relevant**

About the author

Maria Fernanda Reyes Pérez was born in Quito-Ecuador in March 1983. She graduated from Universidad San Francisco de Quito (USFQ), in Quito, Ecuador, in June 2007, with a Bachelor Degree in Environmental Engineering. After her studies, she had the opportunity to work in USFQ in the Masters of Ecology Department, as well as participating in several projects and consultancies in environmental remediation.

In 2011 she obtained her master's degree in Environmental and Energy Management from the University of Twente -The Netherlands. After her master's studies, she taught at USFQ the class of energy and Environment, worked in several consultancy projects and worked in the Ministry of Public Health of Ecuador as environmental engineer.

While working at USFQ, she also contributed in several projects and researches in the Galápagos Islands. Therefore, her motivation to carry out a research in this place was developed, after visiting this place in several occasions. The main purpose of this research was to contribute with the environment, as well as with the local community. In 2012, she started her PhD at UNESCO-IHE in Delft, The Netherlands. Currently, she is a PhD candidate under the department of Environmental Engineering and Water Technology. Her research interest includes water demand management, water supply, water balance, multi-criteria decision analyses, water prognosis, environmental management and sustainability.

List of Publications

Journal Papers

Reyes, M., Trifunović, N., Sharma, S., Behzhadia K., Kapelan, Z., Kennedy, M. (2017) *Mitigation Options for Future Water Scarcity: A Case Study of Santa Cruz Island (Galápagos Archipelago)*. Water. **9** (8): 597.

Reyes, M., Trifunović, N., Sharma, S., Kennedy, M. *(2017). Assessment of domestic consumption in intermittent water supply networks: Case study of Puerto Ayora (Galápagos Islands)***.** Accepted in Journal of Water Supply, Research and Technology (AQUA).

Reyes, M.F., Trifunović, N., D' Ozouville, N., Sharma, S., Kennedy, M. (2017) *Quantification of Urban Water Demand in the Island of Santa Cruz (Galápagos Archipelago).* Journal of Desalination and Water Treatment Journal **64** (1): 1-11

Reyes, M. F., Trifunović, N., Sharma, S., Kennedy, M (2016). *Data assessment for water demand and supply balance on the island of Santa Cruz (Galápagos Islands).* Journal of Desalination and Water Treatment **57**(45): 21335-21349.

Reyes, M., Trifunović, N., Sharma, S , Kennedy, M. (2015) *Water Supply and Demand in Santa Cruz Island-Galápagos Archipelago.* International Water Technology Journal, 6-3 p. 212-221. International Water Technology Association.

Reyes, M., Petricić, A., Trifunović, N., Sharma, S., Kennedy, M. (2017) Multi-Criteria Decision Analysis of Water Demand Management Options for Puerto Ayora (Galápagos Islands). Under review in Water Resources Management Journal.

Other Publications

Reyes, M., Trifunović, N.,Sharma, S. Kennedy, M. (2015) *Water Supply Assessment on Santa Cruz Island: A Technical Overview of Provision and Estimation of Water Demand.* pp. 46-53. In: Galápagos Report 2013-2014. GNPD, GCREG, CDF and GC. Puerto Ayora, Galápagos, Ecuador.

Reyes M., Trifunović, N., Sharma, S., Kennedy, M. (2015) *Evaluación del Suministro de Agua en la Isla Santa Cruz: Una Perspectiva General Técnica Sobre la Provisión y Demanda Valorada de Agua.* Pp. 46-53. En: Informe Galápagos 2013-2014. DPNG, CGREG, FCD y GC. Puerto Ayora, Galápagos, Ecuador.

Reyes, M.,Trifunovic, N.,Sharma, S. Kennedy, M. (2017). *Estimation and Forecasting of Water Demand in Puerto Ayora.* In: Galápagos Report 2015-2016. GNPD, GCREG, CDF and GC. Puerto Ayora, Galápagos, Ecuador.

Reyes, M.,Trifunovic, N.,Sharma, S. Kennedy, M. (2017) *Estimación y Pronóstico de la Demanda de Agua en Puerto Ayora.* En: Galápagos Report 2015-2016. GNPD, GCREG, CDF and GC. Puerto Ayora, Galápagos, Ecuador.

Conference Papers and Oral Presentations

Reyes, M., Trifunović, N., Sharma, S., Kennedy, M. (2015) *Water Supply and Demand in Santa Cruz Island-Galápagos Archipelago*. Manuscript presented at the 18th International Water Technology Conference, Sharm el Sheikh, Egypt, March 12-14, 2015.

Reyes, M., Trifunović, N., Sharma, S., Kennedy, M. (2015) *Implications of Water Tariff Structure on Water Demand in Santa Cruz Island (Galápagos Archipelago)*. Manuscript presented at the XVth World Water Congress, Edinburgh, Scotland, May 24-29, 2015.

Reyes M., Trifunović, N., Sharma, S., Kennedy, M. (2016) *Water Supply and Demand Forecasting in Puerto Ayora-Santa Cruz*. Manuscript presented at the 14th Islands of the World Conference, Lesvos, Greece, May 23-27, 2016.

Reyes M., Petricic A., Trifunović, N., Sharma, S., Kennedy, M. (2017) *Analysis for Water mitigation options using MCDA: A Case Study in the Galápagos Islands*. Manuscript to be presented at the 9th International conference on sustainable Water Resources Management, Prague, Czech Republic, July 18-20, 2017.

Reyes, M., Trifunović, N., Sharma, S., and Kennedy, M. (2017). *Hydraulic modelling of demand growth in tourist islands, case study: Galápagos, Ecuador*. Manuscript presented at the 15th International Symposium on Water Management and Hydraulic Engineering, Primošten, Croatia, September 6-8, 2017.

SENSE

Netherlands Research School for the
Socio-Economic and Natural Sciences of the Environment

D I P L O M A

For specialised PhD training

The Netherlands Research School for the
Socio-Economic and Natural Sciences of the Environment
(SENSE) declares that

Maria Fernanda Reyes Perez

born on 23 March 1983 in Quito, Ecuador

has successfully fulfilled all requirements of the
Educational Programme of SENSE.

Delft, 28 September 2017

the Chairman of the SENSE board

Dr. Ad van Dommelen

the SENSE Director of Education

Prof. dr. Huub Rijnaarts

K O N I N K L I J K E   N E D E R L A N D S E
A K A D E M I E   V A N   W E T E N S C H A P P E N

The SENSE Research School declares that Ms Maria Reyes has successfully fulfilled all
requirements of the Educational PhD Programme of SENSE with a
work load of 41.5 EC, including the following activities:

<u>SENSE PhD Courses</u>

- Environmental research in context (2013)
- SENSE writing week (2014)
- Research in context activity: 'Taking initiative, organising and communicating
  stakeholder meeting on 'Water problems in Santa Cruz: Implications and
  Recommendations' (Santa Cruz, Ecuador - 2015)

<u>Other PhD and Advanced MSc Courses</u>

- Water transport and distribution, UNESCO-IHE (2013)
- Python course, UNESCO-IHE (2014)
- Academic writing course, Premier Taaltraining (2014)
- Modelling and Information systems for water, UNESCO-IHE (2015)
- Presentation skills, Delft University of Technology (2015)
- The art of scientific writing, Delft University of Technology (2016)

<u>Management and Didactic Skills Training</u>

- Supervising two MSc students with thesis entitled 'Water demand forecasting in the
  Galapagos Islands' (2014-2015) and 'Multi-criteria analysis of water supply and demand
  for tourist islands' (2015-2016)

<u>Oral Presentations</u>

- *Water supply and demand in Santa Cruz.* PhD week UNESCO-IHE, 27 September 2014,
  Delft, The Netherlands
- *Water demand and supply in Santa Cruz Island - Galapagos Archipelago.* International
  Water Technology Conference, 12-14 March 2015, Sharm El Sheikh, Egypt
- *Implications of water tariff structure on water demand in Santa Cruz Island (Galápagos
  Archipelago).* World Water Congress XV, 25-29 May 2015, Edinburgh, Scotland
- *Water supply and demand forecasting in Santa Cruz Island: Galápagos Archipelago.* ISISA
  Islands of the World XIV Conference, 23-27 May 2016, Lesbos Island, Greece
- *Analysis for Water Mitigation Options using MCDA: A Case Study in the Galapagos
  Islands.* 9th International Conference on Sustainable Water Resources Management, 18-
  20 July 2017, Prague, Czech Republic

SENSE Coordinator PhD Education

Dr. Ing. Monique Gulickx

Printed and bound by CPI Group (UK) Ltd, Croydon, CR0 4YY

21/10/2024

01777096-0007